FOOD SCIENCE AND TECHNOLOGY

BEEF

PRODUCTION AND MANAGEMENT PRACTICES

FOOD SCIENCE AND TECHNOLOGY

Additional books in this series can be found on Nova's website under the Series tab.

Additional e-books in this series can be found on Nova's website under the eBooks tab.

FOOD SCIENCE AND TECHNOLOGY

BEEF

PRODUCTION AND MANAGEMENT PRACTICES

NELSON ROBERTO FURQUIM
EDITOR

Copyright © 2018 by Nova Science Publishers, Inc.

All rights reserved. No part of this book may be reproduced, stored in a retrieval system or transmitted in any form or by any means: electronic, electrostatic, magnetic, tape, mechanical photocopying, recording or otherwise without the written permission of the Publisher.

We have partnered with Copyright Clearance Center to make it easy for you to obtain permissions to reuse content from this publication. Simply navigate to this publication's page on Nova's website and locate the "Get Permission" button below the title description. This button is linked directly to the title's permission page on copyright.com. Alternatively, you can visit copyright.com and search by title, ISBN, or ISSN.

For further questions about using the service on copyright.com, please contact:
Copyright Clearance Center
Phone: +1-(978) 750-8400   Fax: +1-(978) 750-4470   E-mail: info@copyright.com.

NOTICE TO THE READER

The Publisher has taken reasonable care in the preparation of this book, but makes no expressed or implied warranty of any kind and assumes no responsibility for any errors or omissions. No liability is assumed for incidental or consequential damages in connection with or arising out of information contained in this book. The Publisher shall not be liable for any special, consequential, or exemplary damages resulting, in whole or in part, from the readers' use of, or reliance upon, this material. Any parts of this book based on government reports are so indicated and copyright is claimed for those parts to the extent applicable to compilations of such works.

Independent verification should be sought for any data, advice or recommendations contained in this book. In addition, no responsibility is assumed by the publisher for any injury and/or damage to persons or property arising from any methods, products, instructions, ideas or otherwise contained in this publication.

This publication is designed to provide accurate and authoritative information with regard to the subject matter covered herein. It is sold with the clear understanding that the Publisher is not engaged in rendering legal or any other professional services. If legal or any other expert assistance is required, the services of a competent person should be sought. FROM A DECLARATION OF PARTICIPANTS JOINTLY ADOPTED BY A COMMITTEE OF THE AMERICAN BAR ASSOCIATION AND A COMMITTEE OF PUBLISHERS.

Additional color graphics may be available in the e-book version of this book.

Library of Congress Cataloging-in-Publication Data

ISBN: 978-1-53613-254-0
Library of Congress Control Number: 2018933787

Published by Nova Science Publishers, Inc. † New York

For Marcia, my dear wife.
For all reasons only the two of us know.

CONTENTS

Foreword		**xi**
	Luis Madi	
Introduction		**xv**
Chapter 1	The Supply of a Beef and Liver Hamburger: A High Iron Content Alternative Food	**1**
	Jacqueline Dias Machado de Melo,	
	Natália De Luca Silva,	
	Olivia Borges, Paula Louro Silva,	
	Ana Cristina M. Moreira Cabral	
	and Nelson Roberto Furquim	
Chapter 2	The Effects of Herbs and Spices on the Sensory and Physicochemical Properties of Healthier Emulsified Meat Products: Adding Value by Natural Antioxidant Claims	**35**
	Ana Karoline Ferreira Ignácio Câmara	
	and Marise Aparecida Rodrigues Pollonio	

viii   *Contents*

Chapter 3   The Supply of Safe Beef in Brazil:
                  A Discussion about the Efficacy
                  of the Traceability System Implemented
                  in the Production Chain   **69**
                  Nelson Roberto Furquim

Chapter 4   Exploring the Brazilian Consumer's
                  Perception about Sodium Chloride
                  Reduction in Frankfurter Type Sausages:
                  A Qualitative and Quantitative Approach   **93**
                  Maria T. E. L. Galvão, Rosires Deliza
                  and Marise A. R. Pollonio

Chapter 5   The Relationship of Fatty Acid Composition,
                  Cholesterol Content, Nutritional and
                  Enzymatic Indices with the Intramuscular
                  Fat Content of Beef   **123**
                  Pilar T. Garcia, Nestor N. Latimori,
                  Ana M. Sancho and Jorge J. Casal

Chapter 6   The Brazilian Green Beef:
                  The Importance of Pasture Management for
                  Animal Performance and Quality of the Meat **151**
                  Bruno Lala, Vinícius Valim Pereira,
                  Ulysses Cecato, Guilherme Sicca Lopes Sampaio,
                  Ana Paula Possamai and Ana Maria Bridi

Chapter 7   Nutritional and Genetic Factors That Affect
                  Meat Quality   **175**
                  Otávio Rodrigues Machado Neto,
                  Josiane Fonseca Lage, Liziana Maria Rodrigues
                  and Mateus Silva Ferreira

Chapter 8   Packaging of Red Meat   **193**
                  José Boaventura M. Rodrigues
                  and Claire Isabel G. L. Sarantópoulos

Contents ix

Chapter 9 Pre-Slaughter Cattle Management in Brazil:
Animal Welfare and Meat Quality **225**
Bruna Domeneghetti Smaniotto
and Bruno Lala

About the Editor **247**

List of Contributors **249**

Index **259**

FOREWORD

As presented on the website of the Institute of Food Technology (Instituto de Tecnologia de Alimentos - ITAL) www.alimentos processados.com.br, the beginning of food technology goes back to the prehistory of mankind. With creativity and talent, empirically over the centuries, people have discovered and improved ways to prepare and conserve foods, giving rise to the basis of various processing techniques used in contemporary society. With advances in food science, these techniques have been incorporating scientific principles into the experience thus forming the basis of the food technology currently employed in the manufacture of food ingredients and products as well as the packaging for storing processed foods and beverages.

The Tropical Center for Food Research and Technology (Centro Tropical de Pesquisa e Tecnologia de Alimentos - CTPTA), created in 1963, from the innovative vision of the then governor of the State of São Paulo, became our Institute of Food Technology (ITAL) in 1969, initiating science and food technology in Brazil.

According to the industrial census of 1920, the Brazilian food industry had 2,709 companies out of a total of 13,336 in the whole country. The meat processing industry with 99 companies, or 3.7% of the total, however, presented great potential for growth.

Visualizing this potential, the president of Brazil inaugurated in 1976 the pilot plant of meat processing at ITAL (http://ital.sp.gov.br/50anos). We were at the beginning of science and technology in the meat and meat processing industry in Brazil, currently the flagship of the Brazilian Agribusiness.

In 2016, according to the Brazilian Association of Food Industries (Associação Brasileira das Indústrias da Alimentação-ABIA), the meat and meat processing industry represented 21.7% of the Food and Beverage Industry in Brazil, practically double of the second food sector which was coffee and cereals.

Today, Brazil can be considered the World Champion in the Production of Food and Beverage with Sustainability.

From 1990/91 to 2016/17, according to São Paulo School of Economics of the Getulio Vargas Foundation (FGV / EESP), the area of grain production increased from 38 million to 89 million hectares, a growth of 59%. During this same period grain production increased from 58 million tons to 227.9 million tons, an expansion of 294%.

When analyzing beef production in the same period, according to the Brazilian Association of Animal Proteins (Associação Brasileira de Proteína Animal -ABPA), beef production rose from 5 million tons in 1990 to 9.3 million tons, an increase of 85%.

Data from the Brazilian Beef Exporters Association (Associação Brasileira das Indústrias Exportadoras de Carne - ABIEC) in 2015 show that the value of the production chain was R $ 483.5 billion, with 19.66% of export-oriented production and 81.34% for the domestic market.

This book, *Beef: production and management practices*, which I have the honor to preface, was written by professionals and co-workers of ITAL, the Brazilian Agricultural Research Corporation (Empresa Brasileira de Pesquisa Agropecuária -EMBRAPA), University of Campinas (Universidade Estadual de Campinas-Unicamp), University of São Paulo (USP), São Paulo State University (UNESP), Faculty of Agronomy and Agro-Food Sciences, University of Moron (Argentina), Food Technology Institute (Argentina), INTA Marcos Juarez Experimental Station (Argentina), and Mackenzie Presbyterian University, among other

Foreword

institutions, bringing scientific, technological and market information extremely valuable, showing Brazil as one of the world's biggest exporters of quality beef, and the excellence of beef production in the southern cone of Latin America.

It is important to emphasize the importance of the topics presented as production with traceability, animal welfare, pasture and quality of the meat produced. It also addresses aspects of healthiness, such as salt reduction, use of herbs, nutritional and genetic factors, ending with a fundamental item for this strategic sector which is packaging.

I feel really glad to preface it, for it will contribute to the world market and the area of science and technology, providing very useful information about the meat sector.

Luis Madi
General Director of the Institute of Food Technology (ITAL),
São Paulo State Secretariat of Agriculture and Food Supply

INTRODUCTION

Beef is a nutritionally important food, it is a source of protein, vitamins and minerals, and its production plays an important role in the economy of several countries.

Taking into account its importance as food and the evolution of the educational process of consumers, who have become increasingly well informed and demanding, the demand for quality assured beef with origin certification has increased significantly, as well as the concerns with health and wellness issues.

From the perspective of beef supply, that demand happens both due to the intrinsic quality attributes of the food itself, such as softness, flavor, fat content, as well as characteristics related to the different ways of production, animal breeding and transportation, traceability, environmental issues, processing, marketing, inspection and control, among others.

As there has been an increase in the beef production worldwide and the international issues of trading that food are always dynamic, special attention is required at all stages of the production chain to promote the quality of the finished product, and further to that, several management practices are highlighted.

Challenges and opportunities for beef import and export markets may then rise, with impacts on their economies, at the same time as the requirements imposed by them become more strict and constant.

Quality and safety of products are key elements for the sector, which prioritizes aspects related to animal health, adequate production and processing practices, sanitary and phytosanitary rules and barriers, nutritional concerns, traceability and labeling, as well as changes in the consumer behavior.

Reinforcing its economic and job generating importance, the worldwide beef sector may still be positively impacted by the population increase, the free trade, the increasing improvements in transportation and logistics and the availability of new knowledge and cutting-edge technologies.

Information has always been an important input for agribusiness, both in production and marketing. With the growth in size, competitiveness and, consequently, the complexity of worldwide agribusiness in recent years, knowledge has become an even more essential tool.

Knowledge generated by experimental studies plays a relevant role for the formulation of recommendations on good production practices and management strategies, insofar it may provide the basis for understanding how different agents of a productive chain interact, and how those interactions might benefit and promote the safe supply of food. Thus, this book addresses important contexts in the beef production. Productive aspects are pointed out, focusing on differentiated systems of production (pasture and confinement), and some of the factors (feeding, age, genetic groups, cattle handling, gender) that interfere in the quality of the meat supplied.

Throughout the chapters are also discussed approaches on beef production systems and management, the relevance of bovine farming in the economic context, the use of mechanisms for inspection and control of both animal health and production, and marketing of beef and beef based products.

There is no intention that subjects will be exhausted. The idea is that each one of the articles, shown as chapters of the book, fulfill the role of presenting updated scientific researches, consolidating and demonstrating

Introduction xvii

the work of highly qualified professionals in their respective fields of activity. That may certainly lead to future studies, stimulating the continuous search for significant contributions to the beef sector, which is very promising for the worldwide economy.

Nelson Roberto Furquim, PhD
Editor

In: Beef
Editor: Nelson Roberto Furquim

ISBN: 978-1-53613-254-0
© 2018 Nova Science Publishers, Inc.

Chapter 1

THE SUPPLY OF A BEEF AND LIVER HAMBURGER: A HIGH IRON CONTENT ALTERNATIVE FOOD

Jacqueline Dias Machado de Melo,
Natália De Luca Silva, Olivia Borges,
Paula Louro Silva, Ana Cristina M. Moreira Cabral[*]
and Nelson Roberto Furquim, PhD
Mackenzie Presbyterian University,
São Paulo, SP, Brazil

ABSTRACT

Iron deficiency is a nutritional disorder linked to several physiological and public health problems, such as anemia, individual development, growth, and it may occur in persons of many age groups.

To avoid problems arising from low iron consumption, it is necessary the consumption of food rich in this micronutrient, besides the

---

[*] Corresponding Author Email: ana.cabral@mackenzie.br.

consumption of other foods that help in the absorption of iron. For this study it was proposed the development of a hamburger with two important sources of iron: ground beef and liver, combined with protein ingredients, such as texturized soy protein. For conservation of finished products, freezing technology has been considered. In Brazil, food control and regulation are a shared responsibility between public administration bodies and entities. The Ministry of Agriculture, Livestock and Supply (MAPA) is responsible for the rules and control of slaughterhouses and companies processing livestock based foods. The National Health Surveillance Agency (ANVISA) is responsible for establishing regulatory guidelines aligned with international agencies like the Codex Alimentarius, for packing standards, including food labeling, limits for veterinary drugs and allowed food additives.

Another significant aspect considered in the study was the packaging of the product to be marketed: cardboard box combined with individual plastic film.

A sensory analysis test (preference test) with the finished product was conducted with a convenience, non-probabilistic sample of 61 people, in which each individual analyzed four variables (appearance, taste, odor and texture) on a hedonic scale.

The results obtained indicated that the hamburger was well accepted regarding texture, taste, and odor. It is considered a food rich in protein, low sodium and rich in iron, according to RDC N° 54, November 12, 2012, which provides the Technical Regulation on Complementary Nutrition Information.

Compared to other hamburgers available in the Brazilian market, it was possible to observe that the ground beef and liver based hamburger has lower energy value, less total fat, more protein, and lower sodium content.

It was then concluded that the proposed hamburger might be a product of great nutritional importance for the whole population, mainly for being a source of iron food, feasible as far as production is concerned, being an effective alternative for persons with iron deficiency.

Keywords: liver, beef, hamburger, iron deficiency, sodium

INTRODUCTION

The fortification of food by the addition of nutrients has been utilized by many countries as a measure of public health on the prevention of disabilities in large population segments. Among the advantages observed in the fortification, it is noticed that, although this modifies the nutrients intake, it does not modify the population's eating habits, besides being related to food nutritionally adequate and widely consumed. () Basic food for the population is, however, a suitable carrier to reach the at-risk groups, even allowing the addition of more than one nutrient (Brazil, 2007a).

There are fortification experiences in several countries. With regards to iron, fortifications have already been carried out in various foods, such as: powdered milk, children's formulations, wheat flour, biscuits, among others. The experiments involved different tested carriers, and have shown satisfactory results in the fight against nutritional deficiencies of micronutrients (Brazil, 2007a).

According to the Food Guide for the Brazilian Population (2014), in the last decades Brazil has undergone rapid demographic, epidemiological and nutritional transitions, presenting as consequences: a longer life expectancy, a reduction in the number of children per woman and significant changes in food consumption.

Despite the intense reduction of malnutrition in children, micronutrient deficiencies and chronic malnutrition are still prevalent among vulnerable population groups. There is also a significant increase in overweight and obesity in all age groups.

The changes in the patterns of Brazilian feeding habits involve substitutions of *in natura* foods or minimally processed foods from vegetable origin by instant or ready to eat industrialized products, that bring, among other consequences, imbalance in nutrient supplies and excessive calorie intakes (Brazil, 2014).

It is noticed that the food industry develops options of healthy food products and that meets, among other aspects, the nutritional needs of the population.

The present study was guided by the idea of enriching with iron a beef hamburger, using bovine liver among the main ingredients, considering the positive results of food fortification as a way of facing the prevalence of anemia in a portion of the population. According to Domellof (2014), anemia affects 1.62 billion people, being 293.0 million pre-school children. In Brazil, anemia affects 45.00% of pre-school children, under 5 years old (UNICEF, 2005).

1. THEORETICAL BACKGROUND

1.1. Quality Control of Meat in Brazil

A quality product fits perfectly to the customer's needs which requires to be reliable and affordable. When considering a food product such as beef, and the customer is a modern, demanding and very selective consumer, this definition could include the concepts of nutritional value, health and organoleptic characteristics (Embrapa, 1999).

The attributes of meat quality can be classified as: a) visual quality – aspect that attracts or repels the consumer that goes shopping; b) gustatory quality - attributes that makes the consumer return or not buy the product; c) nutritional quality - nutrients that make the consumer create a favorable or unfavorable image of the meat, as food compatible with their requirements for a healthy life, and d) safety – hygienic-sanitary aspects and the presence or not of chemical contaminants (Embrapa, 1999).

The meat safety aspects permeate the entire production system, from the food provided to the cattle to the packed meat available in the supermarket's shelves. That is why, it is primordial the development of technologies associated with food safety throughout the productive chain, encompassing prevention, detection, early adoption of control measures and eradication of disease, among other pertinent aspects (Embrapa, 2017).

As described by Furquim (2012), Brazil started to have basic legislation on food with the enactment of Decree Law 986 of October 21st, 1969 (Brazil, 1969), laying down basic rules on food and establishing in its Article 3, that mentions: "All food will be exposed to consumption or delivered for sale after being registered with the competent body of the Ministry of Health." In turn, Article 4 defines: "For the grant of registration the competent authority will comply to standards set by the National Commission of Food Standards."

The National Health Surveillance Agency (ANVISA), an autarchy entity created by Law number 9,782 of January 26th, 1999 (Brazil, 1999), is currently the competent Body of the Ministry of Health (MS) for the registration of food and, as required by law, in its Article 7, subsection III, it has competence to "establish norms, propose, monitor and execute policies, guidelines and all actions regarding health surveillance". According to Article 8 of this same law, under the responsibility of ANVISA, "in compliance with current legislation, it regulates and controls products and services which involve a risk to public health".

The industrial and sanitary inspection of products of animal origin in Brazil was regulated in the 1950s, has undergone some adjustments over the years, and is currently ruled by Decree number 2,244, of 1997 (Brazil, 1997). In this intermission, with the Law 7,889, of November 23rd, 1989, of the Ministry of Agriculture, Livestock and Supply - MAPA (Brazil, 1989), the Brazilian sanitary inspection system, until then only under the federal government's responsibility, was restructured and started to contemplate responsibilities at the municipal, state and federal levels, aimed at reducing clandestine animal slaughters, according to Buainain and Batalha (2007b, p. 57).

The Law number 8,171, of January 17th, 1991, amended by Law number 9,712, of 20th November, 1998 (Brazil, 1998), disposed on the agricultural policy and, in its Article 27-A, established that the objectives of agricultural defense are to ensure:

I – the health of plant populations;
II – the health of livestock;
III – the appropriateness of inputs and services used in agriculture;
IV – the identity and hygienic-sanitary and technological safety of finished agricultural products intended for consumers.

Through the following activities to be developed by the Government, according to paragraph 1 of this same article:

I – surveillance and plant health defense;
II – surveillance and animal health defense;
III – inspection and classification of products of plant origin, their derivatives, byproducts and waste of economic value;
IV – inspection and classification of products of animal origin, their derivatives, by-products and waste of economic value;
V – inspection of inputs and services used in agricultural activities.

6   *J. Dias Machado de Melo, N. De Luca Silva, O. Borges et al.*

In this context, it is in force the IN number 10 of 27th April, 2001, of MAPA (Brazil, 2001), that, in its Article 1, prohibits the importation, production, marketing and use of natural or artificial substances with an anabolic activity, or even others endowed with this activity, but devoid of hormone character, for purpose of growth and weight gain in slaughter cattle, with exclusive permission for therapeutic purposes, estrus synchronization, embryo transfer, genetical enhancement and experimental research in veterinary medicine. In fact, the ban on the use of anabolic steroids destined to the increase of bovine weight in Brazil began in 1961 (Ferrão, Bressan, 2006).

Also fall into this scope the programs for the eradication of foot-and-mouth disease, brucellosis and bovine tuberculosis.

The National Program for the Control and Eradication of Brucellosis and Animal Tuberculosis (PNCEBT) was instituted in the country in 2001 by MAPA (Brazil, 2006a), aiming at reducing the impact of these zoonoses on human and animal health, and also to promote the competitiveness of Brazilian livestock. Through this Program, compulsory vaccination against bovine and buffalo brucellosis has been introduced in all geographic areas of the country.

At the same time definition of strategies for certification of regular or monitored properties were put into effect. The National Program for the Eradication and Prevention of Foot-and-Mouth Disease (PNEFA), under the coordination of MAPA, was instituted by IN number 44, of October 2nd, 2007 (Brazil, 2007), which approved the general guidelines (in accordance with the guidance of the World Organization for Animal Health) for the eradication and prevention of foot-and-mouth disease, to be considered throughout the national territory.

The actions of surveillance and sanitary protection of animals and plants would be organized under the coordination of the Public Authorities, in the various federal levels and within their competence, in a Unified System of Attention to Agricultural and Livestock Health (SUASA), according to Article 8 of the Law number 9,712, of 20th November 1998 (Brazil, 1998). SUASA was ruled by Decree number 5,741, of 30th March

2006 (Brazil, 2006b). In its Chapter I, Section I, Article 2, paragraph 3, this decree establishes that:

> Rural producers, industrialists and suppliers of inputs, distributors, cooperatives and associations, industrial and agroindustrial, wholesalers and retailers, importers and exporters, businessman and any other agribusiness operators, along the production chain, are responsible for ensuring that the health and quality of animal and plant products, and of the agricultural inputs are not jeopardized.

Before the consolidation of SUASA, and specifically with regards to the control of herds for the export of beef to markets requiring traceability, aiming at meeting food safety requirements, the Brazilian System of Identification and Certification of Bovine and Buffalo Origin (SISBOV) was established, through the Normative Instruction (IN) number 01, of 9th January 2002, from MAPA (Brazil, 2002a; Sarto, 2002).

The inspection and supervision of plants that produce animal origin goods and that carry out interstate or international trade, are under the responsibility of the Department of Inspection of Products of Animal Origin - DIPOA and of the Federal Inspection Service - SIF, linked to the Ministry of Agriculture, Livestock and Supply.

The inspection and supervision to which this article refers cover, from an industrial and health point of view, the *ante mortem* and *post mortem* inspection of animals, the reception, the handling, the processing, the industrialization, the fractionation, the conservation, the packaging, the packing, the labeling, the storage, the expedition and the transit of any raw materials and products of animal origin (Brazil, 2017a).

The Decree number 9,013 was signed on March 29th, 2017, regulating the Law number 1,283, of 18th December, 1950, and the Law number 7,889, of 23rd November, 1989, which provide the rules for the industrial and sanitary inspection of products of animal origin. The new Regulation of the Industrial and Sanitary Inspection of Products of Animal Origin (RIISPOA) presents among the changes introduced into the legislation, the raising of penalties, new rules to ensure safety and food safety, besides fighting economic fraud.

The revision of RIISPOA contemplates the deployment of new technologies, standardization of technical and administrative procedures, greater harmonization with the international laws, interaction with other public inspection bodies, didactic ordering of norms to facilitate consultation and guidance as well as updating spelling and technical terminologies. The new regulation makes it compulsory to renew the labeling of products of animal origin every 10 years and determines seven types of stamps to be issued by the Federal Inspection Service - S.I.F. (Brazil, 2017b).

The Federal Inspection Service, known worldwide by the initials S.I.F., is linked to the Department of Inspection of Products of Animal Origin - DIPOA, which is responsible for ensuring the quality of edible and inedible animal products intended for the internal and external markets, as well as imported products (Brazil, 2016).

Until it receives the SIF stamp, the product goes through various inspection stages, whose actions are oriented and coordinated by DIPOA, which is part of the Secretariat of Agricultural and Livestock Defense (SDA/MAPA). All products of animal origin under the responsibility of the Ministry of Agriculture, Livestock and Supply are registered and approved by S.I.F. aiming to guarantee products with sanitary and technological certification for the Brazilian consumer, respecting current national and international laws (Brazil, 2016).

1.2. Nutritional Aspects

1.2.1. Anemia

Iron deficiency is the most common nutritional disorder in the world, and it is linked to various problems such as anemia and problems of individual neuropsychomotor development, learning ability, appetite, growth as well as compromising the individual's immune system (Garanito; Pitta; Carneiro, 2010; Domellof, 2014).

The Supply of a Beef and Liver Hamburger

Anemia is divided into three groups: anemia whose erythrocyte production is altered, anemia due to increased erythrocyte malnutrition or anemia due to blood loss (Fabian et al, 2007).

The Iron Deficiency Anemia reaches about 1.62 billion people and 293.0 million children at preschool age in the world, being this group the most affected one (47.40%), being also considered a public health problem, especially in the period of childhood, in which iron requirements are higher.

Adolescence is a complex period in which several factors can directly influence individuals, such as physiological and psychological transformations, sociocultural interferences, unfavorable financial conditions, puberty, among others. There is still an increase in blood volume, of muscle mass and onset of menses in girls, being these factors possible aggravating factors of iron deficiency (Garanito; Pitta; Carneiro, 2010).

Poor eating habits during adolescence is another factor that can cause iron deficiency and possible anemia. Eating habits rich in industrialized foods and fast-food, together with the changes observed in the period of adolescence, can have consequences in the medium and long term in nutritional deficiency of iron (Garanito; Pitta; Carneiro, 2010).

The need for different concentrations of iron in the blood stream changes at each phase of life, and therefore, nutritional recommendations vary with age: for infants between 0-6 months it is 0.27 mg/day, for infants between 7-12 months it is 11.0 mg/day, for children between 1-3 years old it is 7.0 mg/day, for children between 4-8 years old it is 10.0 mg/day, for teenagers between 9-13 years old it is 8.0 mg/day, for teenage boys between 14-18 years old it is 11.0 mg/day and for girls it is 15.0 mg/day (Food and Nutrition Board, 2002).

Iron has important physicochemical properties, mainly in the participation of oxidation and reduction reactions. Iron has a role in the respiratory transport of oxygen and carbon dioxide and is an active part of the enzymes involved in the process of cellular breathing, on immune functions and cognitive development (Czajka-Narins, 1998).

Good sources of iron include meat, fish, poultry, egg yolk, fortified cereals with iron, whole grains, dark green vegetables, beans and peas. Foods with high phytate content such as whole grains and vegetables, have low iron bioavailability, and the tannins of teas, coffee and chocolates also reduce the absorption of iron of vegetable origin, as well as the oxalates found in spinach and beet (Czajka-Narins, 1998).

Ox liver is a nutritious, vitamin-rich food product, and has recently been classified as one of the "superfoods", whose consumption can minimize the incidence of anemia. Among other beneficial compounds present in ox liver are vitamins that can be highlighted such as A, B12, B3, B5, B6 and C. Also found in liver are folate (folic acid), riboflavin, selenium, copper, iron and zinc (Terroni; Bueno; Capobianco, 2015).

1.2.2. Beef

Meat is the set of characteristic tissues with color and consistency, which covers the skeleton of the animals. Commercially, it is named as meat all parts of the animals that serve as food for man, including those from poultry, game, fish and seafood. A cut of meat presents connective tissue, fat and, sometimes, bones (Philippi, 2014).

The meats are sources of protein of high biological value (10.0 to 20.00%), fat (5.0 to 30.00%), vitamins (mainly of complex B: B1, B2, B12 and niacin), vitamin A and minerals (iron, calcium, phosphorus, as well as zinc, magnesium, sodium and potassium). The iron present in the myoglobin of animal origin food is more bioavailable and absorbed (around 15.0 to 35.00%), while iron present in plant origin food is less absorbed, (2.0 to 20.00%), due to its low bioavailability (Philippi, 2014).

Ground beef is a product derived from the grinding of muscle mass of bovine carcasses, followed by immediate cooling or freezing (Brazil, 2003).

1.2.3. Texturized Soy Protein

According to the Resolution of the National Commission of Norms and Standards for Foods (CNNPA) number 14, of June 28th, 1978, that establishes the identity and quality pattern for degreased soy flour,

The Supply of a Beef and Liver Hamburger 11

texturized soy protein (PTS), concentrated soy protein, isolated soy protein and soy extract, PTS is described as:

II – Texturized Soy Protein (PTS)
1. Description
1.1. Definition: Texturized Soy Protein (PTS) is the protein product endowed with identifiable structural integrity, so that each unit supports hydration and cooking, obtained by spinning and thermoplastic extrusion, from one or more of the following raw materials: isolated soy protein, concentrate soy protein and degreased soy flour.
1.1.1. Texturized Soy Protein (PTS) is used as a food ingredient, as a protein source and as an "extender" in meat products. (Brazil, 1978)

PTS can be used itself or added to ground meat (in the proportion of 70.00% of meat to 30.00% of PTS), in the preparation of meatballs and hamburgers (Friedman e Brandon, 2001).

PTS prevents water loss and has a texture similar to meat protein, retains fat, flavors and aromas, forms fiber and adds higher protein value to the finished product. The proof of its benefits to cholesterol control and cardiovascular health encouraged the use of this food in various products such as cereal bars, biscuits and *snacks* (Cassini, 2004).

Daily consumption of 25.0 grams of soy protein can reduce the risk of heart disease, and so the American Heart Association recommends the consumption of foods containing soy for patients with high cholesterol levels (Paschoal, 2001).

In the process of obtaining PTS, care is taken at all stages of the production chain, since planting, storage, preparation, extraction and texturing, aiming to maintain its functional and nutritional properties. PTS is obtained from degreased soy flour, being the extrusion the most important step of the productive process, where the texturing occurs (Cassini, 2004).

The proteins are structural components of the cells, responsible for various chemical reactions and their metabolism, regulating and controlling their external and internal conditions. For food to be considered a good source of protein, bioavailability of essential amino acids must be

12   J. Dias Machado de Melo, N. De Luca Silva, O. Borges et al.

taken into account, besides its digestibility, absence of toxicity and/or presence of antinutritional components (Cassini, 2004).

1.2.4. Sesame

Sesame seeds (*Sesamum indicum L.*) have high nutritional value because of the significant number of vitamins, mainly B complex, and of mineral constituents such as calcium, iron, phosphorus, potassium, magnesium, sodium, zinc and selenium.

The seed is also an important source of edible oil and widely used as seasoning. It has an oil content ranging from 46.00% to 56.00%, with excellent nutritional, medicinal and cosmetic qualities.

Sesame oil is rich in unsaturated fatty acids, such as oleic (47.00%) and linoleic (41.00%), and presents several secondary constituents that are very important in the definition of their chemical properties, such as sesamol, sesamin and sesamoline. Sesamol, with its antioxidant properties, gives the oil high chemical stability, avoiding rancidification, being the one of greater resistance to oxidation among other oils of vegetal origin (Brazil, 2015).

1.3. Food Preservation

When vegetables, fruits, eggs and milk are harvested, when meat is obtained by slaughtered animals, fish taken from its "habitat" and food is prepared, physical, chemical and biological processes begin, which change their organoleptic and health qualities. So, preservation and conservation of food prevails in all stages that precede their consumption, which is reached through various processes, based on partial or total extermination of deteriorating micro-organisms and enzymes, and by the elimination of the predisposing factors (Evagelista, 2008).

Also, according to this author, the methods of food preservation are based on temperature methods (hot and cold); element suppression (water and oxygen); addition of sugar, salt, chemical substances (food additives) and of gases; smoke or use of fermentative agents; freeze-drying process;

food irradiation, among others. To meet the specific needs of each type of food and for conservation to become efficient, there is a need to apply one or more processes, in a combined complementation action.

Freezing is an important and viable method for meats. The main processes utilized for freezing food industrially are: freezing with still air, in plates, with forced circulation of air, by immersion or sprinkling of liquids and cryogenic freezing. The use of low temperatures can control the rate of chemical reactions, that is, the speed at which molecules can move, determining the speed with which they react with each other (Colla, 2003).

When water in the meat is frozen, this annuls almost entirely enzymatic processes and bacterial proliferation (Evagelista, 2008).

Food is frozen taking into account a wide range of temperatures, depending directly on the concentration of salts and water in colloidal suspension in the cells. This way, the freezing speed will depend on the amount of free water present in the cells and also the amount of dissolved salts (Colla, 2003).

The product frozen and thawed according to the recommended standards, will not suffer any reduction of its original features. In order to avoid any loss of organoleptic qualities of the products during the thawing, the parameters of temperature, time, environmental conditions and form of food protection must never be ignored (Evangelista, 2008).

1.4. Packaging

The packaging used in food aims to: protect the food against contamination or losses; facilitate and ensure transport; facilitate the distribution of food; identify the content in terms of quality and quantity parameters and identify the manufacturer. Additionally, it helps to attract the buyer's attention, induce the consumer to buy, instruct him about the use of the product, inform him about the composition, nutritional value and other characteristics of the food, according to legal requirements (Gava, 2008).

The packaging materials should be selected considering a minimum interaction with the food, guaranteeing to the consumers, products that are safe, healthy, and with less preservatives. The challenge of the industries lies in the creation of modern and practical packaging, that preserve food and are environmentally and economically viable (Soares et al., 2009).

Commonly, the materials used in the production of packaging can thus be grouped: rigid metal containers; flexible metal containers; glass; rigid and semi-rigid plastics; flexible plastics; flexible papers; laminated and multifarious; aseptic carton; barrels, cardboard boxes; wood packaging, among others (Gava, 2008).

Depending on their properties, these materials can be used as primary packaging (the one that comes in direct contact with the product), or secondary, tertiary, etc. (which involve other packaging). In this aspect, packaging may also be classified as packaging for transport or retail (Fellows, 2006).

From another perspective, the packaging represents one of the main means of communication between companies and consumers, transmitting visual messages with the purpose of influencing the buying behavior, present nutrition and advertising information (Santesmases Mestre, 1996). In this context, the qualities that most value and exalt the packaging are those that reveal their originality, attractive domain and functional use (Evangelista, 2008).

The choice of food packaging must take into consideration two important aspects: the technological being the first one, in which the packaging must be evaluated due to factors such as mechanical resistance to stacking, transportation, handling and storage; and the second one being public health, where the most relevant contribution of the packaging is the protection of the food against insects, rodents, microorganisms, environmental factors and migration of their own components to the food itself (Fellows, 2006).

Cardboard is a very versatile material used in food packaging: for packaging dried products, as auxiliary packaging and as transportation packaging. As far as fatty foods are concerned, they must be protected by impermeable wrappers or conditioned in a box treated with suitable coatings. Bovine, porcine, poultry and fish meat, both fresh or frozen, can be packed in corrugated boxes, provided that they are appropriate to the characteristics of these products (Evangelista, 2008).

Cardboard box was the type of secondary packaging chosen to pack the hamburger developed in this study, and as its primary packaging, low density polyethylene bags, which present advantages such as: compatibility with various foods, transparency, resistance, easy thermo-welding, water impermeability and low cost (Evangelista, 2008).

2. OBJECTIVES

2.1. Main Objective

The main objective of this study was to develop a hamburger with additional amounts of iron, in order to meet the need of a target audience with a lack of this micronutrient.

2.2. Specific Objectives

- Use bovine liver as a source of iron, and texturized soy protein as a source of vegetable protein to improve the hamburgers texture and emulsion.
- Carry out sensorial tests with the finished product in order to assess the acceptance by potential consumers.
- Conduct a comparative analysis involving 10 brands of beef or chicken hamburgers, available in the Brazilian market, in relation to nutritional aspects informed on the nutritional tables presented in their packaging.

3. METHODOLOGY

For the preparation of beef and liver hamburgers the following ingredients were used: premium beef, liver, ground breadcrumbs, onion, rehydrated texturized soy protein, chicken egg, cumin, herb salt (a blend of light salt, basil, oregano, rosemary), fresh garlic, oregano, black pepper and white sesame.

The following procedure was used to prepare the hamburgers: first all of the formulation ingredients were weighed. Then, a tea cup of texturized soy protein was hydrated in a tea cup of water (200.0 ml) at room temperature and reserved. Also, four liver fillets of approximately 100.0 g each were grounded. In sequence, in a stainless-steel container, the following components of the formulation were homogeneously mixed:

1.0 kg of premium ground beef
4 ground liver fillets (442.0 g)
1 tea cup of hydrated PTS (142.0 g)
1 chopped large onion (144.0 g)
3 teaspoons of herbal salt (14.0 g)
4 teaspoons of cumin powder (20.0 g)
2 teaspoons of black pepper (10.0 g)
2 teaspoons of oregano (10.0 g)
3 chopped garlic cloves (10.0 g)
1 tea cup of ground breadcrumbs (158.0 g)
1 raw chicken egg (53.0 g)
4 tablespoons of white sesame (60.0 g)

The hamburgers were then modeled with their own metallic shapes, in units of 100.0 g each.

For sensory evaluation purposes, they were then roasted in non-stick baking pans, coated with baking paper, in a preheated oven at 180°C, for 40 minutes.

The Supply of a Beef and Liver Hamburger

In order to evaluate the acceptance of the finished products, a test was conducted using the effective sensory method, which assesses the acceptance or preference of consumers for one or more products. In this case, the consumers (without any technical knowledge) responses were used to establish a comparison among competing products and/or in the development of new products (Dutcosky, 2007).

The sensory evaluation was performed on the 4th of May 2017, in the experimental kitchen of Mackenzie Presbyterian University, in the city of São Paulo, with a sample of young people between the ages of 18 and 25, both genders, regular hamburger consumers, weekly and/or monthly mostly. Each individual tried a small sample of the hamburger and was asked to evaluate it in relation to four criteria (appearance, taste, odor and texture) of a hedonic scale, with varying scores between liked it very much (score 9.0) and extremely disliked it (score 1.0). It was also possible to obtain data regarding the frequency of hamburger consumption by the young participants of the tastings sessions.

The data obtained from the sensory evaluations were processed by the researchers, using the Office Package 365 (2016).

The study was approved by the Ethics in Research Committee of Mackenzie Presbyterian University, in the city of São Paulo, Brazil, under the number CAAE – 48483015.7.0000.0084.

4. RESULTS

For the field research on this study 50 individuals were considered, which constituted a non-probabilistic sample.

Table 1 presents data concerning the acceptance of finished products. It is noticed that the highest percentage of the evaluators (32.70%) scored 7 ("Enjoyed regularly") for the appearance attribute. It was observed that 11.47% of the evaluators, scored 4 ("slightly disliked") and 14.70% of the evaluators scored 5 ("Neither liked, nor disliked"). Only 3.27% of the evaluators scored 2 ("Dislike it a lot").

18   *J. Dias Machado de Melo, N. De Luca Silva, O. Borges et al.*

For the odor attribute, 8 ("really enjoyed") was scored by the highest percentage of evaluators (36.00%), and score 7 ("enjoyed regularly") was the second one, concerning 21.30% of the evaluators. For the appearance attribute, score 7 ("enjoyed regularly") was pointed by most evaluators (32.70%). The lowest score given to the appearance attribute was 5 ("neither liked, nor disliked"), corresponding to 13.10% of the evaluators.

When analyzing the data obtained for the flavor attribute, it was possible to observe that 39.34% of the evaluators scored 8 ("really enjoyed"), and only 6.55% scored 4 ("slightly disliked"). Flavor is a mixed sensation, which involves the senses of smell and taste, and yet a set of elements that influence the perception of taste such as: temperature, pressure, astringency, among others (Dutcosky, 2007).

Regarding the texture/softness attribute, the score that most prevailed was 9 ("liked it very much"), corresponding to 37.70% of the evaluators, followed by a score 8 ("really enjoyed"), mentioned by 32.78% of the evaluators.

The results of Table 1 indicate that attributes such as flavor and texture were very well evaluated by most evaluators, with score 8 ("really enjoyed"), by 39.34% of the evaluators and score 9 ("liked it very much"), by 37.70% of the evaluators, respectively.

Table 2 presents the participants frequency of hamburger consumption. According to the data presented, it is observed that 32.70% of the evaluators consume hamburgers monthly and 31.14% consume them weekly. The consumption of hamburger 3 times per week is comparatively smaller, as well as daily consumption. The sporadic consumption of hamburger was pointed out by 22.95% of the consumers.

With regard to the structuring of the nutritional information table for the beef and liver hamburger, it was considered that the roasted hamburger had a loss of 20.0 g in relation to the previous weight, that was 100.0 g. Calculations of nutritional information were performed considering this weight loss.

The Supply of a Beef and Liver Hamburger 19

Table 1. Parameters for acceptance evaluation of beef and liver hamburgers

Evaluation parameters	Attributes			
	Appearance (%)	Odor (%)	Flavor (%)	Texture/ Softness (%)
9.Liked it very much	3.27	16.30	11.47	37.70
8.Really enjoyed	16.30	36.00	39.34	32.78
7.Enjoyed regularly	32.70	21.30	21.30	19.67
6.Liked it slightly	16.30	13.10	18.03	3.27
5.Neither liked, nor disliked	14.70	13.10	3.27	3.27
4. Slightly disliked	11.47	0	6.55	1.60
3.Disliked regularly	1.60	0	0	0
2.Disliked it a lot	3.27	0	0	0
1.Extremelydisliked	0	0	0	1.60
TOTAL	100.00	100.00	100.00	100.00

Source: The authors, 2017.

Table 2. Frequency of hamburger consumption

Frequency	n	%
Never	0	0
Daily	0	0
3x per week	1.0	1.64
2x per week	7.0	11.50
Weekly	19.0	31.14
Monthly	20.0	32.78
Sporadic	14.0	22.95
TOTAL	61.0	100.00

Source: The authors, 2017.

As can be observed from the results in Table 3, the beef and liver hamburger presents 18.0 g of protein, a quantity considered as "high content" (minimum of 12.0 g of protein per portion) by RDC number 54, of November 12th,2012 (Brazil, 2012). Therefore, the studied hamburger can be considered a food with "high content" or "rich in" proteins.

Table 3. Nutrition information for beef and liver hamburger

	Nutritional Information Portion of 80g (1 unit)	
	Quantity per portion (g)	% RDV*
Energetic Value	127 kcal = 534 KJ	6.00
Carbohydrates	9.3	3.00
Proteins	18.0	25.00
Total fat	5.0	9.00
Saturated fat	2.0	9.00
Trans fat	0	0**
Food fiber	1.6	7.00
Iron	0.0073	52.00
Sodium	0.075	3.00

(*) % Daily values of reference based on a diet of 2,000 kcal or 8,400 kJ. The daily values can be higher or lower depending on the energy needs.
**DI not established.
Source: The authors, 2017.

In relation to iron content, for a food to be considered as "source of iron," RDC number 54, of November 12th, 2012 from ANVISA (Brazil, 2012) establishes that it must contain at least 15.00% of the Recommended Daily Intake (IDR) of that mineral in the food portion. To be considered a product with "high content" of iron, the food must have, at least, 30.00% of IDR of iron in the portion. The established IDR for iron is 14.0 mg (Brazil, 2003).

Therefore, the hamburger developed for this study is considered a product with "high iron content" (Anvisa, 2012).

The Supply of a Beef and Liver Hamburger 21

However, although iron has several important properties for humans, it is necessary to watch out for excessive intake of this micronutrient. According to the National Health Surveillance Agency (ANVISA), the Tolerable Upper Intake Level (UL) value established for iron is 40.0 to 45.0 mg/day. It refers to the maximum level of daily intake of a nutrient, biologically tolerable, without risk of adverse health effects, for virtually all individuals.

The hamburger developed for this study also contains low sodium content in its composition. Following the criteria of RDC number 54, of November 12th, 2012 - ANVISA (Brazil, 2012), food is considered with "low sodium content" when it contains, maximum, 80.0mg of sodium per portion and greater than 30.0g or 30.0ml. Just like iron, sodium also has a UL value of daily intake which is 2.3g (Brazil, 2012).

The comparison among different hamburgers available in the Brazilian market, regarding nutritional values is presented in Table 4.

Table 4. Nutritional comparison among different brands of hamburgers in the Brazilian market

Hamburger Brands	Energetic value (kcal)	Protein (g)	Total fat (g)	Sodium (mg)
Proposed forrmulation	127.0	18.0	5.0	75.0
Brand 1	225.0	15.0	17.5	465.0
Brand 2	145.0	12.0	10.0	554.0
Brand 3	160.0	13.0	11.0	629.0
Brand 4	351.0	31.0	25.0	63.0
Brand 5	161.0	12.0	12.0	624.0
Brand 6	181.0	13.0	13.0	535.0
Brand 7	187.0	12.0	15.0	783.0
Brand 8	330.0	30.0	22.5	757.5
Brand 9	177.0	14.0	12.0	757.0
Brand 10	165.0	14.0	11.0	774.0

Source: The authors, 2017.

Analyzing comparatively the nutritional aspects of the beef and liver hamburger for this study, with 10 brands of similar products available in the Brazilian market, it has been observed that concerning brands 1, 2 and 3 (Table 4), it presents lower energy value, less total fat, more protein, and less sodium. Brand 4, however, shows a higher amount of protein, 31.0g; and less sodium, 63.0mg, than the hamburger in this study, even though it has a higher total fat content.

As far as brands 5, 6, 7, 9 and 10 are concerned they present higher energy value, lower amount of protein, higher total fat and sodium content than the hamburger developed in this study. On the other hand, brand 8, as brand 4, present higher amounts of protein and higher energy values.

5. DISCUSSION

It is a fact that, currently, with the market movements, there is an increasing supply of food products. Concomitantly, with the evolution of digital media, the consumers are increasingly given access to information and become more demanding and judicious in their choices.

On the other hand, the food industry seeks to meet not only the demands of conventional consumers, but also of those concerned with health, good shape and the ones with food restrictions. Thus, the supply of food products that meet these specific nutritional demands is growing (Embrapa, 1999).

Numerous factors may interfere in the way food is consumed and also in the lifestyle of consumers, for example urbanization, industrialization of markets, or even population growth, among others. With regard to aspects of public health, and with an increasing prevalence in several populations, inadequate or insufficient intake of nutrients can lead to obesity, diabetes, hypertension, anemia and some types of cancer (Abreu et al., 2001). Therefore, the supply of food products with differentiated nutritional content may appear as a form of contribution to minimize these health hazards.

The Supply of a Beef and Liver Hamburger 23

Bearing in mind that beef is a food which provides consumers with proteins of high biological value besides minerals, iron in particular, and vitamins (Philippi, 2014), in this study it was chosen to develop a food item (hamburger) based on beef and liver, as an alternative option for anemic consumers or those in need of iron-rich diets.

Brazil is currently one of the main worldwide players in the production and trade of beef, due to a structured development process which not only increased productivity but also the quality of Brazilian beef and, consequently, its competitiveness and market share. In 2015, Brazil reached the largest cattle herd (209 million heads), to be the second largest consumer (38.6 kg/inhabitant/year) and the second largest exporter (1.9 million tons carcass equivalent) of beef in the world, with more than 39 million slaughtered cattle (Embrapa, 2017).

As far as health aspects are concerned, the Brazilian cattle industry has built a solid prevention and control structure for the main problems that can either lead to losses in productivity, or risks to consumers health, due to the intense performance of the official health defense agencies and the science and technology institutions (Embrapa, 2017).

Engaged in the search for productivity, quality and sustainability, science and technology institutions, schools, industries, producer associations, non-governmental organizations, among others, make up an extremely active group, with initiatives that greatly contribute to the quality of the supplied beef. Concerning meat quality, the livestock activity is increasingly encouraged to pay attention to the demands of the consumers, either by actions of the slaughterhouses or by the government initiatives (Embrapa, 2017).

Therefore, management practices are implemented both in the farms and in the slaughterhouses, as well as in retail stores, to provide end consumers with food of assured quality (Embrapa, 1999).

With regards to the supply of beef and beef based products, these management practices take into account laws and regulatory mechanisms that aim at industrial and sanitary inspection of products of animal origin, besides ensuring the quality of the finished products (Brazil, 2017a, 1969, 1997).

For the hamburger developed in this study, it was chosen to freeze as a conservation process, due to its technological benefits and convenience for commercialization (Gava, 2008).

As far as the nutritional benefits of the proposed food are concerned, it may be considered an adequate alternative for individuals lacking iron-rich diets, because it is a food with high iron content, (Brazil, 2012), and at the same time 1 presenting both flavor and texture positively perceived by the evaluators (Dutcosky, 2007). It also presents a low sodium content (Brazil, 2012).

Those are important differentials for the commercialization of the finished products, because they can be used as marketing appeals, besides being significant information for potential consumers (Gava, 2008).

It is also important to make an adequate management of all the stages of this proposed food productive chain, considering mechanisms of inspection and control (Embrapa, 2017), associated with leading technologies, that will guarantee to the consumers the supply of products with quality excellence (Gava, 2008).

CONCLUSION

The hamburger developed for this study presents a very differentiated positive nutritional composition compared to similar products available in the Brazilian market. Therefore, it may be considered as a feasible alternative for consumers with iron deficiency.

Because it has lower energy value, less total fat, more protein, and a low sodium content, it can also be considered as a product with significant nutritional value for the general population, in particular children and adolescents, who may eventually find it difficult to eat properly and need good sources of iron in their diet.

It must be highlighted the importance of sanitary controls in all stages of the food production chain, from obtaining the raw materials, up to the finished product supplied to consumers, assuring them the supply of food consistent with manufacturing and quality standards.

The Supply of a Beef and Liver Hamburger

REFERENCES

Abreu, E. S. et al. Alimentação mundial: uma reflexão sobre a história. *Saude soc.* [Worldwide Food Consumption: a reflection on history. *Health soc.*], São Paulo, v. 10, n. 2, p. 3-14, Dec. 2001. Disponível em: <http://www.scielo.br/scielo.php?script=sci_arttext&pid=S0104-12902001000200002&lng=en&nrm=iso>. Acesso em: 06 Set. 2017.

Brasil. Decreto-Lei Nº 986, de 21 de outubro de 1969. *Institui normas básicas sobre alimentos* [Decree-Law Nº 986, of 21st October 1969. Brasília, 1969. Establishes basic food standards]. Available in: <http://www.planalto.gov.br/ccivil_03/decreto-lei/Del0986.htm>. Acesso em: 29 ago. 2017.

Brasil. *Agência Nacional de Vigilância Sanitária. Resolução Comissão Nacional de Normas e Padrões para Alimentos (CNNPA)* n° 14, de 28 de junho de 1978. Em conformidade com o disposto no Capítulo V, artigo 28, do Decreto-Lei n° 986, de 21 de outubro de 1969, a CNNPA resolveu estabelecer o padrão de identidade e qualidade para farinha desengordurada de soja, proteína texturizada de soja, proteína concentrada de soja, proteína isolada de soja e extrato de soja [National Health Surveillance Agency. Resolution of the National Commission of Norms and Standards for Foods (CNNPA) n° 14, of 28th June 1978. In accordance with the provisions of Chapter V, article 28, of Decree-Law n° 986, of 21st October 1969, CNNPA decided to establish the identity and quality standard for degreased soy flour, texturized soy protein, concentrated soy protein, isolated soy protein and soy extract]. Available in: <http://www.anvisa.gov.br/anvisalegis/resol/14_78. htm #>. Acesso em: 11 set. 2017.

Brasil. Lei n° 7.889, de 23 de novembro de 1989. *Dispõe sobre inspeção sanitária e industrial dos produtos de origem animal, e dá outras providências* [Law n° 7.889, of 23rd November 1989. Provides for industrial and sanitary inspection of products of animal origin, and establishes other measures]. Available in: <http://www.planalto.gov. br/ccivil_03/leis/L7889.htm >. Acesso em: 29 ago. 2017.

26 *J. Dias Machado de Melo, N. De Luca Silva, O. Borges et al.*

Brasil. Decreto N° 2.244, de 4 de junho de 1997. Altera dispositivos do Decreto n° 30.691, de 29 de março de 1952, que aprovou o Regulamento da Inspeção Industrial e Sanitária de Produtos de Origem Animal, alterado pelos Decretos n° 1.255, de 25 de junho de 1962, n° 1.236, de 2 de setembro de 1994, e n° 1.812, de 8 de fevereiro de 1996. *Diário Oficial da República Federativa do Brasil, Brasília*, DF, 5 jun. 1997. Seção 1, p. 3. [Decree N° 2.244, of 4th June 1997. Changes devices of Decree N° 30.691, of 29th March, 1952, that approved the Regulation for Industrial and Sanitary Inspection of Products of Animal Origin, amended by Decrees n° 1.255, of 25th June, 1962, n° 1.236, of 2nd September, 1994, and n° 1.812, of 8th February, 1996. Official Journal of the Federative Republic of Brazil, Brasília, DF, 5 jun. 1997. Section 1, p. 3].

Brasil. *Lei n° 9.712, de 20 de novembro de 1998. Altera a Lei n° 8.171, de 17 de janeiro de 1991, acrescentando-lhe dispositivos referentes à defesa agropecuária.* Brasília, 1998 [Law n° 9.712, of 20th November 1998. Amends the Law n° 8.171, of 17th January 1991, adding provisions related to agricultural defense. Brasília, 1998]. Available in: <http://www. planalto.gov.br/ccivil_03/leis/L9712.htm >. Acesso em: 29 ago. 2017.

Brasil. *Lei N° 9.782, de 26 de janeiro de 1999. Define o Sistema Nacional de Vigilância Sanitária, cria a Agência Nacional de Vigilância Sanitária, e dá outras providências* [Law n° 9.782, of 26th January 1999. Defines the National Health Surveillance System, creates the National Health Surveillance Agency, and gives other measures]. Available in: <https://www.planalto.gov.br/ccivil_03/leis/l9782. htm>. Acesso em: 09 abr. 2017.

Brasil. *Ministério da Agricultura e do Abastecimento. Instrução Normativa n° 10, de 27 abril de 2001* [Ministry of Agriculture and Supply. Normative Instruction n° 10, of April 27, 2001]. Available in: <http://sistemasweb.agricultura.gov.br/sislegis/action/detalhaAto.do?m ethod=visualizarAtoPortalMapa&chave=507130696>. Acesso em: 9 abr. 2017.

The Supply of a Beef and Liver Hamburger 27

Brasil. Ministério da Agricultura, Pecuária e Abastecimento. Instrução Normativa Conjunta n° 1, de 09 de janeiro de 2002. Institui o Sistema Brasileiro de Identificação e Certificação de Origem Bovina e Bubalina - SISBOV. *Diário Oficial da República Federativa do Brasil* [Ministry of Agriculture, Livestock and Supply. Joint Regulatory Instruction No. 1, of January 9, 2002. Establishes the Brazilian System of Identification and Certification of Bovine Origin and Bubalina - SISBOV. Official Gazette of the Federative Republic of Brazil], Brasília, DF, 10 jan. 2002. Seção 1, p. 6.

Brasil. Ministério da Agricultura, Pecuária e Abastecimento. Instrução Normativa n° 83, de 21 de novembro de 2003. *Regulamentos Técnicos de Identidade e Qualidade de carne bovina em conserva (Corned beef) e carne moída de bovino* [Ministry of Agriculture, Livestock and Supply. Normative Instruction n° 83, of 21st November 2003. Technical Regulations of Identity and Quality of Preserved Beef (Corned beef) and ground beef]. Available in: < http://www.aladi. org/nsfaladi/normasTecnicas.nsf/09267198f1324b6403257496006234 3c/4207980b27b39cf903257a0d0045429a/$FILE/IN%20N%C2%BA %2083-2003.pdf>. Acesso em: 29 ago. 2017.

Brasil. Ministério da Saúde. Agência Nacional de Vigilância Sanitária. Resolução n° 360, de 23 de dezembro de 2003. *Regulamento técnico sobre rotulagem nutricional de alimentos embalados* [Ministry of Health. National Health Surveillance Agency. Resolution n° 360, of 23rd December 2003. Technical regulation on nutritional labeling of packaged foods]. Available in: <http://portal.anvisa.gov.br /documents/33880/2568070/res0360_23_12_2003.pdf/5d4fc713-9c66-4512-b3c1-afee57e7d9bc>. Acesso em: 29 ago. 2017.

Brasil. *Ministério da Agricultura, Pecuária e Abastecimento. Programa Nacional de Controle e Erradicação da Brucelose e da Tuberculose Animal (PNCEBT)* [Ministry of Agriculture, Livestock and Supply. National Program for the Control and Eradication of Brucellosis and Animal Tuberculosis (PNCEBT)]. Organizado por Vera Cecília Ferreira de Figueiredo, José Ricardo Lôbo e Vitor Salvador Picão Gonçalves. Brasília: MAPA/SDA/DSA, 2006a. Available in:

28 *J. Dias Machado de Melo, N. De Luca Silva, O. Borges et al.*

<http://www>. agricultura.gov.br/arq_editor/file/Aniamal/programa%20 nacional%20sanidade%20brucelose/Manual%20do%20PNCEBT%20-%20Original .pdf>. Acesso em: 06 fev. 2012.

Brasil. *Decreto N° 5.741, de 30 e março de 2006b. Regulamenta os arts. 27-A, 28-A e 29-A da Lei n° 8.171, de 17 de janeiro de 1991, organiza o Sistema Unificado de Atenção à Sanidade Agropecuária, e dá outras providências* [Decree N° 5.741, of 30th March, 2006b. Regulates articles 27-A, 28-A and 29-A of the Law n° 8.171, of 17th January 1991, organizes the Unified System of Attention to Agricultural and Livestock Health, and establishes other measures]. Available in: <http://www.planalto.gov.br/ccivil_03 /_ato2004-2006/2006/decreto /d5741.htm>. Acesso em: 29 ago. 2017.

Brasil. *Ministério da Saúde. Unicef. Cadernos de Atenção Básica: Carências de Micronutrientes/Ministério da Saúde, UNICEF; Bethsáida de Abreu Soares Schmitz. - Brasília: Ministério da Saúde, 2007a. 60 p. - Série A. Normas e Manuais Técnicos* [Ministry of Health. UNICEF. Reviews of Basic Attention: Micronutrient Deficiencies/Ministry of Health, UNICEF; Bethsáida de Abreu Soares Schmitz. - Brasília: Ministry of Health, 2007a. 60 p. - Serie A. Norms and Technical Handbooks]. Available in: <http://www. sbp.com. br/ fileadmin/user_upload/pdfs/Cadernos_Micronutrientes_MS.pdf>. Acesso em: 12 set. 2017.

Brasil. Ministério da Agricultura, Pecuária e Abastecimento. *Instrução Normativa N° 44*, de 02 de outubro de 2007b. Aprova as diretrizes gerais para a Erradicação e a Prevenção da Febre Aftosa, constante do Anexo I, e os Anexos II, III e IV, desta Instrução Normativa, a serem observados em todo o Território Nacional, com vistas à implementação do Programa Nacional de Erradicação e Prevenção da Febre Aftosa (PNEFA), conforme o estabelecido pelo Sistema Unificado de Atenção à Sanidade Agropecuária . [Ministry of Agriculture, Livestock and Supply. Normative Instruction N° 44, of 2nd October, 2007b. Approves the general guidelines for the Eradication and the Prevention of Foot-and-Mouth Disease, in the Attachment I, and the Attachments II, III and IV, of this Normative Instructions, to be observed

The Supply of a Beef and Liver Hamburger 29

throughout the National Territory, aiming at the implementation of the National Program for the Eradication and Prevention of Foot-and-Mouth Disease (PNEFA), as established by the Unified System of Attention to Agricultural and Livestock Health]. Available in: <file:///C:/Users/Ana/Downloads/Instru%C3%A7%C3%A3o%20Nor mativa%20MAPA%20n%C2%BA%2044,%20de%2002%20de%20out ubro%20de%202007%20(1).pdf>. Acesso em: 29 fev. 2017.

Brasil. Ministério da Saúde. Agência Nacional de Vigilância Sanitária. *Resolução n° 54, de 12 de novembro de 2012.* Regulamento técnico sobre informação nutricional complementar [Ministry of Health. National Health Surveillance Agency. Resolution n° 54, of 12th November 2012. Technical regulation about complementary nutritional information]. Available in: <http://portal.anvisa.gov.br/documents/ 33880/2568070/rdc0054_12_ 11_2012.pdf/c5ac23fd-974e-4f2c-9fbc-48f7e0a31864>. Acesso em: 22 ago. 2017.

Brasil. Ministério da Saúde. Secretaria de atenção à saúde. Departamento de Atenção Básica. *Guia alimentar para a população brasileira* [Ministry of Health. Department of Health Care. Department of Basic Care. Food guide for the Brazilian population], 2. ed. Brasília, 2014. Available in: < http://portalarquivos.saude.gov.br/images/pdf/2014/ novembro/05/Guia-Alimentar-para-a-pop-brasiliera-Miolo-PDF-Internet.pdf>. Acesso em: 31 ago. 2017.

Brasil. Ministério da Saúde. Secretaria de Atenção à Saúde. Departamento de Atenção Básica. *Alimentos regionais brasileiros. Brasília: Ministério da Saúde* [Ministry of Health. Department of Health Care. Department of Basic Attention. Brazilian regional foods. Brasília: Ministry of Health], 2. ed. 2015. 484 p.:il. Available in: <http://189.28.128.100/dab/docs/portaldab/publicacoes/livro_alimento s_regionais_brasileiros.pdf> Acesso em: 29 ago. 2017.

Brasil. *Ministério da Agricultura, Pecuária e Abastecimento. Serviço de Inspeção Federal (SIF)* [Ministry of Agriculture, Livestock and Supply. Federal Inspection Service (SIF)]. Available in: < http://www.agricultura. gov.br/assuntos/inspecao/produtos-animal/sif>. Acesso em 22 ago.2017.

30 *J. Dias Machado de Melo, N. De Luca Silva, O. Borges et al.*

Brasil. Decreto n° 9.013, de 29 de março de 2017a. Regulamenta a Lei n° 1.283, de 18 de dezembro de 1950, e a Lei n° 7.889, de 23 de novembro de 1989, que dispõem sobre a inspeção industrial e sanitária de produtos de origem animal. *Diário Oficial da União, Brasília* [Decree n° 9.013, of 29th March 2017. Regulates the Law n° 1.283, of 18th December 1950, and the Law n° 7.889, of 23rd November 1989, which deals with the industrial and sanitary inspection of products of animal origin. Official Diary of the Union, Brasília], DF, 30 março 2017. Seção 1, p. 3. Available in: < http://www2. camara. leg.br/legin/fed /decret/2017/decreto-9013-29-marco-2017-784536-publicacaooriginal-152253-pe.html>. Acesso em: 11 set. 2017.

Brasil. Ministério da Agricultura, Pecuária e Abastecimento. *Novo regulamento da inspeção de produtos de origem animal prevê penas mais severas* [Ministry of Agriculture, Livestock and Supply. New regulation of the inspection of products of animal origin establishes more severe penalties]. RIISPOA. Brasília, DF, 30 mar 2017b. Available in: <http://www.agricultura.gov.br/noticias/novo-regula mento-da-inspecao-de-produtos-de-origem-animal-reforca-seguranca-alimentar>. Acesso em: 29 ago. 2017.

Buainain, A. M.; Batalha, M. O. (Coords.). *Cadeia produtiva da carne bovina* [*Beef productive chain*]. Brasília: IICA/MAPA/SPA, 2007.

Cassini, A. S. *"Análise das características de secagem da proteína texturizada de soja"*. 2004. 136 f. Monografia (Especialização) - Curso de Engenharia Química, Pós Graduação em Engenharia Química ["Analysis of the drying characteristics of texturized soy protein". 2004. 136 f. Term Paper (Specialization) – Chemical Engineering Course, Postgraduate in Chemical Engineering], Universidade do Rio Grande do Sul, Porto Alegre, 2004. Available in: <https://www.lume .ufrgs.br/bitstream/handle/10183/4737/000414 520.pdf?sequence=1>. Acesso em: 22 ago. 2017.

Colla, L. M. *"Congelamento e descongelamento – sua influência sobre os alimentos"*. 2003. 14 f. Monografia (Especialização) - Curso de Engenharia de Alimentos ["Freezing and thawing – its influence on foods". 2003. 14 f. Term Paper (Specialization) – Food Engineering

The Supply of a Beef and Liver Hamburger 31

Course], Universidade Federal do Rio Grande, Rio Grande, 2003. Available in: <http://repositorio.furg.br/bitstream /handle/1/6803 /428-742-1-PB.pdf?sequence=1>. Acesso em: 22 ago. 2017.

Czajka-Narins, D. M. In: *Vitaminas.* Mahan, L. K & Escott, S. S. Alimentos, nutrição e dietoterapia [Food, nutrition and diet therapy]. 9 ed. São Paulo: Roca, 1998. Cap 6. p. 77-122.

Domellof, M. et al. *Iron Requirements of Infants and Toddlers. Espghan Committee on Nutrition,* [s.i.], v. 58, n. 1, p.119-129, Jan. 2014.

Dutcosky, S.D. *Análise sensorial de alimentos [Food sensory analysis].* 2. ed. Curitiba, PR: Champagnat, 2007.

Embrapa (Empresa Brasileira de Pesquisa Agropecuária). Curso conhecendo a carne que você consome 1. Qualidade da carne bovina. *Campo Grande: Embrapa Gado de Corte,* 1999. 25p. (Embrapa Gado de Corte. Documentos, 77). [(Brazilian Agricultural Research Corporation). Course knowing the meet that you eat 1. Quality of the beef. *Campo Grande: Embrapa Cattle,* 1999. 25p. (Embrapa Cattle. Documents, 77)].

Embrapa (Empresa Brasileira de Pesquisa Agropecuária). Nota Técnica. *Evolução e Qualidade da Pecuária Brasileira* [(Brazilian Agricultural Research Corporation). Technical Note. *Evolution and Quality of Brazilian Livestock*]. Rodrigo da Costa Gomes1 Gelson Luiz Dias Feijó2 Lucimara Chiari Campo Grande, 2017. Available in: <https://www.embrapa.br/documents/10180 /21470602/Evolucaoe QualidadePecuaria.pdf/64e8985a-5c7c-b83e-ba2d-168ffaa762ad>. Acesso em: 29 ago. 2017.

Evangelista, José. *Tecnologia de Alimentos [Food Technology].* São Paulo: Editora Atheneu, 2008.

Fabian, C; Olinto, A. T. M; Costa, J. S. D. Prevalência de anemia e fatores associados em mulheres adultas residentes em São Leopoldo, Rio Grande do Sul [Prevalence of anemia and associated factors in adult women living in São Leopoldo, Rio Grande do Sul]. *Cad. Saúde pública.* vol. 23, n. 5, março, 2007.

32 *J. Dias Machado de Melo, N. De Luca Silva, O. Borges et al.*

Fellows, P. J. *Tecnologia do Processamento de Alimentos - Princípios e Prática* [*Food Processing Technology - Principles and Practice*]. 2. ed. Porto Alegre: Artmed, 2006.

Ferrão, S. P. B.; Bressan, M. C. O uso de agentes anabolizantes na produção de carnes e suas implicações: revisão [The use of anabolic agents in meat production and its implications: revision]. *Vet. Not. Uberlândia*, vol. 12, nº 1, p. 69-78, 2006.

Food and Nutrition Board, Institute of Medicine-National Academy of Sciences. *Dietary Reference Intakes*: *Recommended levels for individual intake*. 2002.

Friedman, M.; Brandon, D. L. Nutritional and health benefits of soy proteins. *Journal of Agriculture and Food Chemistry*, v. 49, n. 3, p. 1069-1086, 2001.

Furquim, N. R. "*Alimento seguro: uma análise do ambiente institucional para oferta de carne bovina no Brasil.*" Tese de Doutorado ["Safe food: an analysis of the institutional environment for beef supply in Brazil." Doctoral Thesis]. Universidade de São Paulo, 2012.

Garanito, M. P.; Pitta, T. S.; Carneiro, J. D. A. Deficiência de ferro na adolescência. *Revista Brasileira de Hematologia e Hemoterapia* [Iron deficiency in adolescence. *Brazilian Journal of Hematology and Hemotherapy*], São Paulo, v. 1, n. 1, p. 1-4, 15 Jan. 2010.

Gava, A. J.; Silva, C. A. B.; Frias, J. R. G. *Tecnologia de alimentos*: *princípios e aplicações*. [*Food technology*: *principles and applications*]. São Paulo: Nobel, 2008.

Paschoal, V. *Alimentos para a saúde. Revi. sadia light*, São Paulo, v. 1, pp. 16-17. Dez 2001. Presidência da República Casa Civil Subchefia para Assuntos Jurídicos. Decreto nº 9.013, de 29 de março de 2017. Regulamenta a Lei nº 1.283, de 18 de dezembro de 1950, e a Lei nº 7.889, de 23 de novembro de 1989, que dispõem sobre a inspeção industrial e sanitária de produtos de origem animal [Food for health. Revi. sadia light, São Paulo, v. 1, pp. 16-17. Dec 2001. Republic Presidency Civil House Legal Sub-Office. Decree nº 9.013, of 29th march, 2017. Regulates the Law nº 1.283, of 18th December 1950, and the Law nº 7.889, of 23rd November 1989, which deals with the

The Supply of a Beef and Liver Hamburger 33

industrial and sanitary inspection of products of animal origin]. Available in: <http://www.planalto.gov.br/ccivil_03/_ato2015-2018 /2017/decreto/D9013.htm>. Acesso em: 22 ago. 2017.

Philippi, S. T. *Nutrição e técnica dietética* [*Nutrition and dietary technique*]. 3. ed. Barueri: Manole, 2014.

Santesmases Mestre, M. *Términos de marketing* [*Marketing terms*]. Madrid: Pirámide, 1996.

Sarto, F. M. *"Análise dos impactos econômicos e sociais na implementação da rastreabilidade na pecuária bovina nacional. Trabalho de conclusão de curso"* (Graduação em Engenharia Agronômica) ["Analysis of the economic and social impacts in the implementation of traceability for national cattle". Term Paper (Graduation in Agronomic Engineering)], Universidade de São Paulo, Escola Superior de Agricultura "Luiz de Queiroz", Piracicaba, 2002.

Soares, N. F. F. et al. Novos desenvolvimentos e aplicações em embalagens de alimentos [New developments and applications in food packaging]. *Revista Ceres*, Viçosa, MG v. 56, n. 4, p.370-378, jul. 2009. Mensal. Available in: <www.ceres.ufv.br/ ojs/index.php/ceres /article/view/3438/1341>. Acesso em: 22 ago. 2017.

Terroni, H. C.; Bueno, S. M.; Capobianco, M. P. *Encapsulação de Fígado Liofilizado* [*Freeze dried Liver Encapsulation*]. 2015. Available in: <http://www.unilago.edu.br/revista/edicaoatual/Sumario/2015/ downloads/ 8.pdf>. Acesso em: 22 ago. 2017.

UNICEF. Vitamin & Mineral Deficiency. *A Global Progress Report*. 2005.

In: Beef
Editor: Nelson Roberto Furquim

ISBN: 978-1-53613-254-0
© 2018 Nova Science Publishers, Inc.

Chapter 2

THE EFFECTS OF HERBS AND SPICES ON THE SENSORY AND PHYSICOCHEMICAL PROPERTIES OF HEALTHIER EMULSIFIED MEAT PRODUCTS: ADDING VALUE BY NATURAL ANTIOXIDANT CLAIMS

Ana Karoline Ferreira Ignácio Câmara[*]
and Marise Aparecida Rodrigues Pollonio, PhD

Department of Food Technology, Faculty of Food Engineering,
University of Campinas (UNICAMP),
Cidade Universitária Zeferino Vaz, Campinas, São Paulo, Brazil

ABSTRACT

Consumers are increasingly concerned about healthy eating. Despite the high nutritional value of meat and meat products, their consumption

---

[*] Corresponding Author Email: anakarolineign@gmail.com/pollonio@unicamp.br.

36   A. K. Ferreira Ignácio Câmara and M. A. Rodrigues Pollonio

has been widely criticized due to their high levels of sodium, additives, and mainly saturated fats. Thus, several strategies have been used to reduce or partially replace pork back fat, including the modification of the fatty acid profile in emulsified meat products, reducing the saturated fatty acids content (SFA) and increasing the polyunsaturated fatty acids (PUFAs), especially the ω-3 fatty acids. However, the aspects related to the food safety and shelf-life of emulsified meat products should be fully understood to assess the suitability of these strategies for a healthier reformulation. In addition, such strategies can lead to the production of higher added value products, with a significant increase in the beef supply chain. One of the major challenges of this reformulation is the increase in lipid oxidation rates, which may result in losses in the sensory characteristics and nutritional value of the processed products. Many herbs and spices, commonly used to provide flavor and aroma to foods, contain various phytochemicals that are sources of natural antioxidants, such as phenolic acids, flavonoids, and terpenoids. This study investigated the effects of nine herbs and spices in different blends (rosemary, coriander, white and Jamaica pepper, marjoram, sage, and thyme) on the sensory, physicochemical, and nutritional properties of bologna sausage made with the partial replacement of pork fat by linseed oil (LO) during storage. Five treatments were prepared with the replacement of pork fat by LO (54%) and the addition of herb and spice blends (BHS). A formulation without the addition of BHS (FC1, 20% fat) and a formulation containing linseed oil (FC2) was used as a control. The treatments with LO and BHS were stable to lipid oxidation during the shelf life of the products, and a higher antioxidant protection was observed in the treatments containing thyme (F4 and F5). The lipid reformulation led to an increase of more than 70% of PUFAs (polyunsaturated fatty acids) when compared to FC1, with a good sensory acceptance, except for the attribute color. The results demonstrated the viability to produce bologna sausages with a better nutritional profile. Herbs and spices are sources of natural antioxidants, and have proven to be a viable alternative to replace some food additives and to add value to meat products.

Keywords: claim label, linseed oil, fatty acids, natural antioxidant, herbs and spices, lipid oxidation

1. INTRODUCTION

The last decades have seen a rapid expansion of knowledge about the role of diet on health and well-being of individuals. Despite the nutritional value of meat and meat products as a source of high biological value proteins, lipids, B vitamins, and minerals, such as iron and zinc (Bohrer, 2017), their consumption has been widely criticized due to the high levels of sodium, additives, and saturated fat. Several studies have reported the association between high consumption of these compounds and the increased risk of chronic diseases, particularly hypertension, cardiovascular diseases and various types of cancer (Aggarwal *et al.*, 2017; Wood *et al.*, 2008), which has led the Public Health Agencies to make several recommendations, including the reduction of a saturated fat intake, which should not exceed 10% of the total dietary lipids daily (WHO, 2003).

Many of the negative connotations associated with the processing of meat products can be overcome by reducing the unhealthy constituents such as saturated fats, sodium, and additives. The use of bioactive ingredients in the formulations can reduce such constituents and result in functional meat products, which can bring health benefits when consumed as part of a balanced diet, as well as opening up new markets for the meat processing industry (Grasso *et al.*, 2014).

In general, meat products are characterized by high levels of saturated fat since one of the most important ingredients is back fat, which gives excellent technological improvements to the meat systems, despite being rich in saturated fatty acids. In comminuted meat products, such as bologna sausages, fat plays a very important role in determining the texture, flavor, and physicochemical stability of the product. Traditionally, fats come from adipose animal tissues directly incorporated into meat batters as ingredients. The four major factors that determine the firmness of the adipose tissue are water, lipid content, the amount of connective tissue, and the fatty acid composition. In particular, the origin of the fatty acids plays a prominent role in determining the consistency of these tissues, and the melting point of fatty acids are strongly influenced by the degree of unsaturation of the hydrocarbon chain (Lebret *et al.*, 1996).

Although several strategies have been studied to reduce or partially replace the saturated fats in meat products, from a technological and sensory point of view, it is important to consider the major challenge of these reformulations, once the high melting point of saturated fats plays important functional properties related to the batter stability (Delgado-Pando *et al.*, 2010; Selani *et al.*, 2016). Changes in the fatty acid profiles of meat and meat products can be obtained by feeding diets rich in PUFAs (polyunsaturated fatty acids), including fish oil and linseed oil (Raes *et al.*, 2004; Lu *et al.*, 2008) or by the incorporation of these oils as ingredients in the manufacturing of meat products (Alejandre *et al.*, 2017; Ansorena & Astiasarán, 2004).

Bologna sausages represent one of the most widely consumed meat products in Brazil, with significant production volumes, and contain up to 30% fat, as established by the current legislation (Brasil, 2000). Thus, the reformulation of these products is required, and in this context many studies have reported the modification of the fatty acid profile in meat products, with a reduction of saturated fatty acids (SFA) and an increase in PUFAs, especially the ω-3 fatty acids (Jung & Joo, 2013, Gallardo *et al.*, 2015, Câmara & Pollonio, 2015). Some authors have reported the importance of a balanced intake of ω-6 to ω-3 fatty acids to promote health benefits, with a recommended a n-6/n-3 ratio of 5: 1 or less (Schaefer, 2002; Simopoulos, 2004; Bernardi *et al.*, 2016).

The addition of vegetable oils such as olive, linseed, and sunflower oils in emulsified meat products has been suggested for this purpose; however, many technological limitations related to the product's stability need to be overcome for its application on an industrial scale (Valencia *et al.*, 2008; Salcedo-Sandoval *et al.*, 2015, Jiménez-Colmenero *et al.*, 2013, Xiong *et al.*, 2016). One of the major challenges is the increase in the lipid oxidation reactions due to the higher MUFAs and PUFAs levels, which can negatively affect the sensory characteristics and the nutritional value of the processed products (Negre-Salvayre *et al.*, 2008; Jiang & Xiong, 2016).

The prevention of lipid oxidation or the reduction of the reaction rate is a prerequisite for the successful development of PUFA-rich meat products (Jacobsen *et al.*, 2008). Although synthetic antioxidants such as butylated hydroxytoluene (BHT), butylated hydroxyanisole (BHA) and propyl gallate (PG) have been used for a long time to inhibit the harmful changes induced by the oxidation of meat and meat products, studies have shown their potential genotoxic effects (Dolatabadi & Kashanian, 2010; Verhagen *et al.*, 1991). In addition, natural antioxidants such as herbs and spices meet the current industrial trend of producing foods with cleaner labels.

Many herbs and spices commonly used to provide flavor and aroma to foods contain various phytochemicals that are sources of natural antioxidants, such as phenolic acids, flavonoids, terpenoids, tannins, and others (Shan *et al.*, 2005; Vallverdú-Queralt *et al.*, 2014). Studies have shown the high antioxidant activity of rosemary, sage, oregano, marjoram, and thyme, herbs from the Labiatae family (Zheng & Wang, 2001; Houssain *et al.*, 2011). It was also found high antioxidant capacity in clove, cinnamon, and black pepper (Przygodzka *et al.*, 2014; Slavin *et al.*, 2016) in the form of extracts.

Erdmann *et al.* (2017) evaluated the oxidative stability of emulsified meat products enriched with 1% fish oil and 50 mg/kg of encapsulated rosemary extract and found an effective antioxidant protection during 21 days of refrigerated storage (7 °C) when compared to the control. Trindade *et al.* (2010) reported that irradiated beef burgers containing rosemary (400 mg/kg) and oregano (400 mg/kg) extracts, alone or in combination with BHA/BHT (200 mg/kg), led to a decrease in lipid oxidation in the samples stored at -20 °C for 90 days.

Other studies have reported the effective protection of herbs and spices as natural antioxidants in meat products (Hernández *et al.*, 2009; Kong *et al.*, 2010; Jinap *et al.*, 2015). However, few studies use combinations of different herbs and spices to evaluate the synergistic antioxidant effects of these compounds used in the dehydrated form (a more usual form of addition) in the reformulated meat products. An additional challenge refers to the development of very distinct flavors that can affect the meat matrix, changing the sensory characteristics of a particular meat product.

40   A. K. Ferreira Ignácio Câmara and M. A. Rodrigues Pollonio

In this study, nine herbs and spices (rosemary, coriander, marjoram, oregano, white pepper, Jamaica pepper, parsley, sage, and thyme) were selected due to their recognized antioxidant activity and frequent use in the meat industry as aromatic ingredients. The antioxidant capacity of these herbs and spices in bologna sausages with lipid profile modified with linseed oil was investigated through an accelerated oxidation test, and those with the higher antioxidant protection were selected to make the blends. The main objective of the study was to evaluate the effects of the herb and spice blends on the physicochemical, nutritional, and sensory characteristics of bologna sausage reformulated with linseed oil during the product's shelf life.

2. MATERIALS AND METHODS

2.1. Materials

Bologna sausages were made using lean meat (*M. trapezius cervicis*, 73.96% moisture, 3.01% pork fat) and pork (*Triceps brachii*, 72.22% moisture and 4.10% pork fat) purchased from an industrial supplier (JBS SA, Brazil). Fat and apparent aponeuroses were removed before processing. Meat and the pork fat were ground on a 3.5 mm disk.

Linseed oil (Vital Âtman, Brazil) was obtained by the cold extraction method, and used to prepare the oil in water emulsion (fatty acid composition: myristic (0.05), pentadecanoic (0.03), palmitic (5.03), palmitoleic (0.06), margaric (12.06), margaroleic (0.05), stearic (3.77), oleic (23.28), linoleic (17.06), t-linolenic acid (0.21), linolenic (49.78), arachidic (12.15), gadoleic (0.16), behenic (0.15) and lignoceric (0.16)). Sodium caseinate (96% protein, Global Foods, São Paulo, Brazil) was selected as a protein base for preparing the pre-emulsion.

The other ingredients and additives were sodium chloride (Cisne, SA Rio de Janeiro, Brazil), starch (Corn Products, São Paulo, Brazil), garlic and onion powder (Ibrac, São Paulo, Brazil), sodium tripolyphosphate, sodium erythorbate, and sodium nitrite, provided by ISP Germinal, São

Paulo, Brazil. The herbs and spices were rosemary, coriander, marjoram, oregano, white pepper, Jamaica pepper, parsley, sage, and thyme (Fuchs Gewurze, Itupeva, Brasil).

2.2. Selection of Herbs and Spices with Antioxidant Properties

Nine herbs and spices were selected (as described in *Section 2.1*) and added (1%) separately to the meat batters with 54% replacement of pork fat by linseed oil. The substitution level of pork fat by linseed oil, as well as the technological viability of the process, were established in a previous study (Câmara & Pollonio, 2015). Two control treatments without the addition of herbs and spices were performed, one containing only pork fat (20%) and the other with 54% replacement of pork fat by linseed oil, totaling eleven treatments. The processing of bologna sausages was carried out according to the procedures described in *Section 2.5*.

After cooling, the bologna sausages were sliced and subjected to an accelerated oxidation test and placed in expanded polystyrene (EPS) trays (in triplicate for each sample) covered with oxygen permeable packaging in a cold room (7 °C), lighting with constant 3000 to 3500 LUX. The samples were collected for five days, every 24 hours, and frozen for determination of TBARS (thiobarbituric acid reactive substances).

2.3. Mix of Herbs and Spices Blends

According to the results of the accelerated oxidation test, the herb and spice blends were elaborated, with the herbs and spices with the lowest oxidation profile, as determined in *Section 2.2, Selection of herbs and spices with antioxidant properties*, besides considering the attributes flavor and aroma. Five different blends were made, and added to the bologna sausages formulations at a concentration of 0.5%. The composition of the blends is shown in Figure 1.

42 A. K. Ferreira Ignácio Câmara and M. A. Rodrigues Pollonio

BLEND 1	BLEND 2	BLEND 3	BLEND 4	BLEND 5
Rosemary, Coriander, White pepper, Jamaica pepper.	Rosemary, Coriander, White pepper, Jamaica pepper, Marjoram.	Rosemary, Coriander, White pepper, Jamaica pepper, Sage.	Rosemary, Coriander, White pepper, Jamaica pepper, Thyme.	Rosemary, Coriander, White pepper, Jamaica pepper, Marjoram, Thyme, Sage.

Figure 1. Composition of blends of herbs and spices used in the bologna sausages formulations with the replacement of pork fat by linseed oil.

Table 1. Percentages of linseed oil, pork fat, herbs, and spices used in the bologna sausages formulations

| Ingredients | Treatments (%) | | | | | | |
	FC1 (Control)	FC2 (Control with oil)	F1	F2	F3	F4	F5
Beef	27	27	27	27	27	27	27
Pork	27	27	27	27	27	27	27
Pork backfat	20	7.2	7.2	7.2	7.2	7.2	7.2
Linseed oil	0	10.8	10.8	10.8	10.8	10.8	10.8
Sodium caseinate (SC total)	*2*	*2*	*2*	*2*	*2*	*2*	*2*
(SC) in the pre-emulsion	*0*	*1.08*	*1.08*	*1.08*	*1.08*	*1.08*	*1.08*
(SC) in the batter	*2*	*0.92*	*0.92*	*0.92*	*0.92*	*0.92*	*0.92*
Ice (I total)	*18.08*	*20.08*	*19.58*	*19.58*	*19.58*	*19.58*	*19.58*
(I) in the pre-emulsion	*0*	*8.64*	*8.64*	*8.64*	*8.64*	*8.64*	*8.64*
(I) in the batter	*18.08*	*11.44*	*10.94*	*10.94*	*10.94*	*10.94*	*10.94*
Rosemary	*0*	*0*	*0.125*	*0.1*	*0.1*	*0.1*	*0.071*
Coriander	*0*	*0*	*0.125*	*0.1*	*0.1*	*0.1*	*0.072*
White pepper	*0*	*0*	*0.125*	*0.1*	*0.1*	*0.1*	*0.072*
Jamaica pepper	*0*	*0*	*0.125*	*0.1*	*0.1*	*0.1*	*0.072*
Marjoram	*0*	*0*	*0*	*0.1*	*0*	*0*	*0.071*
Sage	*0*	*0*	*0*	*0*	*0.1*	*0*	*0.071*
Thyme	*0*	*0*	*0*	*0*	*0*	*0.1*	*0.071*

2.4. Treatments and Experimental Design

Six bologna sausages formulations with modified lipid profile were made with 54% replacement of pork fat by pre-emulsified linseed oil. Of these, five were named F1, F2, F3, F4, and F5, and the herb and spice blends were added (Fig. 1). Two treatments without the addition of herbs and spices were used as a control: (FC1) containing 20% pork fat, and (FC2) with modified lipid profile, as shown in Table 1. The following ingredients and/or additives were also used (%) in each treatment: starch, 3.3; condiments, 0.255; sodium tripolyphosphate, 0.2; sodium erythorbate, 0.05; sodium nitrite, 0.015; and sodium chloride, 2.00.

2.5. Bologna Sausage Processing

The linseed oil was pre-emulsified according to the procedure described by Paneras *et al.* (1998). For that, eight parts of warm water (37 °C) were mixed for 2 minutes with one part of sodium caseinate until total dissolution, and then the mixture was emulsified in ultra turrax (IKA T25 digital, Staufen, Germany) for 3 minutes with ten parts of oil. This procedure was performed the day before. Table 1 shows the treatments, and the amount of water and sodium caseinate used in the linseed oil pre-emulsion. Refrigerated meat and pork fat (1 to 2 °C) were milled using 3.5 mm orifice disks, and the emulsion was processed into a cutter (Mado®), as follows: the ingredients meat, salt, half the ice, sodium nitrite, sodium tripolyphosphate, and seasonings were comminuted until the temperature reached 7 °C. Then, the remaining ice was added along with the sodium erythorbate and the pork fat and/or oil-in-water pre-emulsion (depending on the treatment). After that, sodium caseinate was added, and the comminution was carried out at a temperature not exceeding 12 °C. The meat batter (0.5 kg) was embedded in plastic casings of 70-75 mm in diameter (Spel Embalagens, São Paulo, Brazil) in a filling machine (Mainca®), and cooked in an oven (Arprotec®, Brazil) according to the following cooking schedule: 15 minutes at 65 °C and 98% relative humidity (RH), 20 minutes at 75 °C and 98% RH, and then 85 °C and 98% RH until the internal temperature reached 72 °C. The products were then cooled in cold potable water and stored in a cold room (4 °C) until analysis. About 150 g raw meat batter was sampled from each treatment, placed in plastic tubes and kept at refrigeration temperature (4 ºC) to perform the emulsion stability test. The experiment was carried out in triplicate.

2.6. Physicochemical Characterization

The moisture, protein, and ash levels were determined according to the methodology described by the Association of Official Analytical Chemists (AOAC, 2005). The fat content was quantified by the method described by Bligh & Dyer (1959). All determinations were performed in triplicate at 24h after processing.

The emulsion stability test was performed according to Hughes *et al.* (1997), with some modifications. Approximately 25 g raw meat batter of each treatment was placed in plastic tubes and centrifuged for 1 minute at 3600 rpm. Soon after, the samples were heated in a water bath at 70 ºC for 30 minutes and centrifuged again for 3 minutes at 3600 rpm. The supernatant was placed in 50 mL beakers, and oven dried at 100 ºC for 16 hours. All determinations were performed in quadruplicate, at 24 h after processing.

Color measurements were performed using a pre-calibrated Hunter Lab (Colourquest-II Hunter Associates Laboratory Inc., Virginia, USA) colorimeter operating with D65 illuminant, 20-mm specimen size, 10° observer angle, in RSEX mode in CIELAB color system, where L* represents the luminosity, a* represents the red-green axis, and b* the yellow-blue axis. The color variables were measured at four points in the central part of the cut surface of four slices per sample. The assays were performed in triplicate for each treatment. The samples were evaluated 24 h after processing and every 15 days during 60 days of shelf-life.

Texture profile analysis was performed at room temperature in TA-xT2i texture analyzer (Texture Technologies Corp., Scarsdale, NY), using six bologna sausages cylinders 20 mm in diameter and 20 mm height for each treatment. The samples were compressed at 30% of their original weight, using test, pre-test, and post-test speeds of 1.00 mm/s and force of 0.10 N, and a P-35 probe (long shaft/normal base). The parameters evaluated were firmness (N), cohesiveness, and chewiness. The samples were evaluated at 24 h after processing and every 15 days during 60 days of shelf-life.

The thiobarbituric acid reactive substances (TBARS) were determined in 2.5 g sample, according to the method proposed by Salih *et al.* (1987), with some modifications. The samples were kept in an ice bath and homogenized in ultra turrax for 1 minute with 10 mL 5% trichloroacetic acid solution (TCA) and 10 mg BHT. The mixture was centrifuged at 3000 rpm for 10 min and the supernatant was filtered. Three milliliters of the filtrate were mixed with 3 mL of 0.02 M 2-thiobarbituric acid (TBA), and heated at 95 °C for 60 min. After cooling, the absorbance was measured at 532 nm in a UV-visible spectrophotometer. The results were expressed as malonaldehyde (MDA) per kilogram of sample (mg/kg) using a standard calibration curve of 1,1,3,3-tetraethoxypropane (TEP). All determinations were performed in quadruplicate at 0, 15, 30, 45, and 60 days.

2.7. Fatty Acid Analysis

The total lipids of the samples were extracted by the method of Bligh & Dyer (1959), and methylation was carried out according to AOCS (2004) standards. The samples were analyzed in CGC Agilent 6850 Series GC Capillary Gas Chromatograph, equipped with DB-23 AGILENT (50% cyanopropyl-methylpolysiloxane) capillary column 60 m, Ø int: 0.25 mm, 0.25 μm film. The operating conditions were: column flow = 1.00 mL / min; linear velocity = 24 cm / sec; detector temperature: 280 °C; injector temperature: 250 °C; oven temperature: 110 °C - 5 minutes, 110 – 215 °C (5 °C / min), 215 °C - 34 minutes; carrier gas: helium; volume injected: 1.0 μL; split ratio: 1:50. The fatty acid composition was determined by comparing the peak retention times with those of the respective fatty acid standards. For each treatment, the measurements were made in duplicate at 0 and 60 days.

46   *A. K. Ferreira Ignácio Câmara and M. A. Rodrigues Pollonio*

2.8. Sensory Evaluation

The sensory evaluation was approved by the Research Ethics Committee of the University of Campinas SP, Brazil (N. 281/2010). A hedonic test was used to evaluate the acceptance of the products for the attributes color, aroma, flavor, texture, and overall impression. For that, 78 participants were recruited among students, staff, and professors at the campus. The samples were served in a monadic form in a balanced, complete block in plastic dishes coded with three-digit numbers, in individual cabins. A 9-point structured hedonic scale was used, with extremes ranging from "disliked very much" to "liked very much " (Stone & Sidel, 1993).

2.9. Statistical Analysis

A completely randomized experimental design was used, with three independent replicates. Data were analyzed by analysis of variance (ANOVA). The means were compared by the Tukey's test at a confidence level of 5% (P <0.05), using the statistical program SAS, version 9.1 (SAS, 2002).

3. RESULTS AND DISCUSSION

3.1. Selection of Herbs and Spices with Antioxidant Properties

The antioxidant properties of the members of the *Lamiaceae* or *Labiatae* family are due to the presence of several phenolic compounds, including rosmarinic acid and caffeic acid in rosemary, sage, oregano, and thyme (Exarchou *et al.*, 2002); flavonoids and rosmarinic acid in marjoram (Ninfali *et al.*, 2005); volatile oils and phenolic amides in the white pepper of the *Piperaceae* family (Shan *et al.*, 2005); thymol and eugenol in the

The Effects of Herbs and Spices on the Sensory ... 47

jamaica pepper of the *Myrtaceae* family (Uhl, 2000); gallic acid in the coriander and flavonoids in the parsley, both members of the *Umbelliferae* family (Wangensteen *et al.*, 2004), among many other compounds with antioxidant properties.

However, in the accelerated oxidation test, not all herbs and spices exhibited a good antioxidant performance during the five days of evaluation. It is worth mentioning that the herbs and spices used in the present study were in the dehydrated form (the simplest form of addition in processing), which may have led to eventual losses of the antioxidant activity during the dehydration process. According to the results in Figure 2, the TBARS values ranged from 0.001 to 0.6 mg MDA / kg, except for parsley (P), which presented a progressive increase, with higher values when compared to the other herbs, reaching TBARS value of 1.15 mg MDA / kg. The detection limit of lipid oxidation in meat products suggested by Gray & Pearson (1987) is 1.0 mg MDA / kg sample.

Although the literature has shown the antioxidant power of herbs such as parsley, the mechanisms of lipid oxidation reactions and antioxidant mechanisms in multiphase food systems, such as the meat batters enriched with vegetable oils, are extremely complex. Many factors can influence the rate and extent of lipid oxidation, making it difficult to predict the antioxidant behavior and effectiveness in these systems. Lipid oxidation in emulsions can occur in all phases, and important intrinsic factors such as the transition metal ions can differentiate the meat emulsions from other simpler systems such as oils, which makes them more complex (Jacobsen *et al.*, 2008).

In addition, some authors have reported the pro-oxidant activity of some herbs under certain circumstances. Wong & Kitts (2006) evaluated the antioxidant and antibacterial capacity of parsley and coriander extracts, and found a pro-oxidant activity in the aqueous extracts of these herbs. According to the authors, although the reducing capacity of the compounds can protect the senescence of parsley and coriander leaves, the pro-oxidant potential of an herb can be due to the same reducing capacity in the presence of free transition metals, especially iron in the case of meat products. This property of the herbs keeps the metal in its reduced and active form, which is necessary to initiate lipid oxidation.

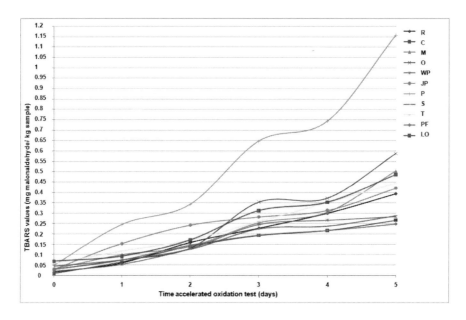

Figure 2. TBARS values of the accelerated oxidation test of the bologna sausages made with the replacement of pork fat by linseed oil and herbs and spices. All treatments have 7.2% pork fat and 10.8% linseed oil, except PF (20% pork fat), R, Rosemary; C, Coriander; M, Marjoram; O, Oregano; WP, White Pepper; JP, Jamaica Pepper; P, Parsley; S, Sage; T, Thyme; LO, Linseed Oil. PF and LO: without the addition of herbs and spices.

In relation to the antioxidant protection of the other herbs evaluated in this study (Figure 2), it was observed that coriander (C), sage (S), and white pepper (WP) presented lower TBARS values at the end of the study, when compared to treatment containing linseed oil without the addition of herbs (LO) and very close to the value found for the treatment containing only pork fat (PF), with favorable antioxidant effects. In contrast, the treatments with rosemary (R), Jamaica pepper (JP) and thyme (T) presented less antioxidant protection, despite the lower TBARS values when compared to the treatment LO. Regarding the treatments with marjoram (M) and oregano (O), similar behavior was observed during the

The Effects of Herbs and Spices on the Sensory ... 49

accelerated oxidation test, once TBARS values were higher than the control treatment with linseed oil (LO) at the end of the test (day 5).

Karpinska *et al.* (2001) evaluated the antioxidant capacity of sage and a blend made with sage, red and black pepper, garlic, and marjoram added to meat cakes made with mechanically separated turkey meat, fried in soybean oil, and stored at refrigeration temperature for 4 days. The addition of sage and the herb and spice blends delayed the oxidation process, and salvia proved to be more effective when compared to the blend.

3.2. Physicochemical Analysis

Bologna sausages were made with five different blends (Figure 1), according to the previous results. The physicochemical characterization was determined for all treatments during the shelf-life, as shown in Table 1 (five blends, two controls, one pork fat, and one modified lipid profile).

The proximate composition is presented in Table 2. The moisture content of the treatments ranged from 61.18% $\pm$ 0.25 to 64.33% $\pm$ 0.09 in FC1 (Control) and F2, respectively. It was observed that the control FC1 (20% pork fat) with a lower moisture content had a higher fat content (17.48%) and differed from all other treatments (P <0.05), probably due to the balance of the formulations since the lower amount of fat was replaced by ice in this treatment.

The emulsion stability is one of the most important parameters in the lipid reformulation of meat products, with a direct effect on the sensory characteristics and the shelf life of the processed products. According to Table 2, all treatments presented good emulsion stability, with very similar values. These results may be due to both the noble formulation used in this study composed of 54% meat raw materials, and the use of the pre-emulsified linseed oil stabilized with sodium caseinate. This protein ingredient has emulsifying properties which together with the meat proteins, covers the fat particles and produces a very strong bond between

50   *A. K. Ferreira Ignácio Câmara and M. A. Rodrigues Pollonio*

the emulsion constituents, resulting in a more compact and stable meat matrix (Cáceres *et al.*, 2008).

Table 2. Proximate composition and emulsion stability of the bologna sausages made with different levels of linseed oil, pork fat, herbs, and spices

Treatments	Moisture (%)	Fat (%)	Protein (%)	Ash (%)	Water released (%)	Fat released (%)
FC1	61.18 ± 0.25^d	17.48 ± 0.42^a	15.02 ± 0.11^a	2.94 ± 0.05^a	1.22 ± 0.06^b	0.092 ± 0.08^b
FC2	64.14 ± 0.10^{ab}	16.23 ± 0.29^b	14.38 ± 0.27^{abc}	2.68 ± 0.07^{bc}	1.57 ± 0.23^{ab}	0.105 ± 0.017^{ab}
F1	63.97 ± 0.03^{ab}	16.04 ± 0.38^b	14.20 ± 0.22^c	2.69 ± 0.01^{bc}	1.60 ± 0.30^{ab}	0.105 ± 0.018^{ab}
F2	64.33 ± 0.09^a	15.07 ± 0.06^c	14.25 ± 0.27^{bc}	2.74 ± 0.02^b	1.55 ± 0.10^b	0.098 ± 0.012^b
F3	64.15 ± 0.02^{ab}	15.85 ± 0.14^{bc}	14.79 ± 0.26^{abc}	2.54 ± 0.12^c	1.35 ± 0.19^b	0.087 ± 0.012^b
F4	63.00 ± 0.09^c	16.46 ± 0.12^b	14.75 ± 0.18^{abc}	2.58 ± 0.02^c	1.83 ± 0.24^{ab}	0.125 ± 0.011^{ab}
F5	63.87 ± 0.18^b	16.04 ± 0.46^b	14.91 ± 0.13^{ab}	2.55 ± 0.06^c	2.18 ± 0.28^a	0.141 ± 0.017^a

* Values represent the average ± standard deviation; All treatments, except FC1 (20% pork fat), have 7.2% of pork fat and 10.8% of linseed oil. [a,b,c] Means in the same column with the same letters did not differ significantly at P < 0.05. FC1: Control (20% pork fat); FC2: Control with linseed oil. R: Rosemary; JP: Jamaica Pepper; WP: White Pepper; C: Coriander; M: Marjoram; S: Sage; T: Thyme. F1: 0.5% of blend 1 (R, JP, WP and C); F2: 0.5% of blend 2 (R, JP, WP, C and M); F3: 0.5% of blend 3 (R, JP, WP, C and S); F4: 0.5% of blend 4 (R, JP, WP, C and T); F5: 0.5% of blend 5 (R, JP, WP, C, M, S and T).

The results of the color measurements during storage are presented in Table 3. The L* values (luminosity) were significantly different among the treatments (P <0.05). All treatments containing linseed oil exhibited higher L* values when compared to the control with the addition of pork fat, FC1, probably due to the milky appearance of the pre-emulsion. In relation to the storage time, a certain stability in the luminosity values was observed during storage, with no significant changes at the beginning and end of the shelf life for this parameter. Regarding the coordinate a*, the FC1 (control) differed (P <0.05) from all other formulations, presenting higher values during the entire storage time. López-López *et al.* (2009) reported an increase in luminosity and a reduction in red color of sausages made with the replacement of pork fat by olive oil. Concerning the b* values, significantly lower values were observed for FC1 when compared to the other treatments (P <0.05). In general, the herb and spice blends impacted in the yellow tonality, in a similar way for all treatments.

The Effects of Herbs and Spices on the Sensory ...

Table 3. L* values (luminosity), a* values (red color intensity), b* values (yellow color intensity) of the bologna sausages made with different levels of linseed oil, pork fat, herbs and spices during storage

	0 day	15 days	30 days	45 days	60 days
L*					
FC1	68.78 ± 0.12^{dB}	68.92 ± 0.24^{eAB}	68.17 ± 0.24^{dC}	69.59 ± 0.16^{dA}	68.78 ± 0.26^{dB}
FC2	72.46 ± 0.07^{aB}	73.10 ± 0.12^{aAB}	71.46 ± 0.22^{aC}	72.83 ± 0.11^{aA}	72.46 ± 0.24^{aB}
F1	72.22 ± 0.47^{aB}	71.46 ± 0.16^{bAB}	71.35 ± 0.25^{aC}	72.45 ± 0.34^{abA}	72.22 ± 0.35^{aB}
F2	70.88 ± 0.61^{cB}	70.94 ± 0.17^{cdAB}	69.94 ± 0.38^{cC}	71.61 ± 0.27^{cA}	70.88 ± 0.18^{cB}
F3	71.53 ± 0.46^{bB}	71.28 ± 0.34^{bcAB}	70.38 ± 0.16^{bcC}	71.47 ± 0.26^{cA}	71.53 ± 0.18^{bB}
F4	70.78 ± 0.44^{cB}	71.76 ± 0.31^{bAB}	70.65 ± 0.24^{bC}	72.23 ± 0.12^{bA}	70.78 ± 0.20^{cB}
F5	71.30 ± 0.53^{bcB}	70.78 ± 0.25^{dAB}	69.89 ± 0.32^{cC}	71.33 ± 0.50^{cA}	71.30 ± 0.38^{bcB}
a*					
FC1	10.81 ± 0.26^{aC}	11.22 ± 0.13^{aB}	11.24 ± 0.07^{aA}	11.14 ± 0.18^{aB}	11.16 ± 0.05^{aAB}
FC2	9.15 ± 0.07^{bC}	9.37 ± 0.10^{bB}	9.69 ± 0.17^{bA}	9.42 ± 0.27^{bB}	9.39 ± 0.33^{bAB}
F1	8.31 ± 0.10^{cC}	8.56 ± 0.04^{cB}	8.58 ± 0.08^{deA}	8.63 ± 0.13^{cB}	8.64 ± 0.08^{cAB}
F2	7.98 ± 0.13^{cdC}	8.67 ± 0.12^{cB}	8.83 ± 0.11^{cA}	8.48 ± 0.12^{cB}	8.43 ± 0.19^{cdAB}
F3	7.89 ± 0.25^{dC}	8.53 ± 0.11^{cB}	8.69 ± 0.06^{cdA}	8.60 ± 0.06^{cB}	8.59 ± 0.11^{cAB}
F4	7.79 ± 0.15^{dC}	8.10 ± 0.14^{dB}	8.39 ± 0.05^{eA}	8.10 ± 0.08^{dB}	8.37 ± 0.12^{cdAB}
F5	7.69 ± 0.08^{dC}	8.16 ± 0.21^{dB}	8.52 ± 0.09^{deA}	8.14 ± 0.13^{dB}	8.13 ± 0.06^{dAB}
b*					
FC1	11.67 ± 0.06^{dC}	12.47 ± 0.10^{dB}	12.83 ± 0.20^{bAB}	12.85 ± 0.14^{dAB}	12.95 ± 0.10^{dA}
FC2	12.71 ± 0.12^{bcC}	13.16 ± 0.23^{cB}	13.33 ± 0.04^{abAB}	13.29 ± 0.18^{cAB}	13.42 ± 0.30^{cA}
F1	13.00 ± 0.04^{aC}	13.24 ± 0.15^{bcB}	13.34 ± 0.45^{aAB}	13.37 ± 0.10^{bcAB}	13.52 ± 0.06^{bcA}
F2	12.83 ± 0.08^{abcC}	13.53 ± 0.07^{aB}	13.68 ± 0.23^{aAB}	13.65 ± 0.22^{abAB}	13.81 ± 0.31^{abcA}
F3	12.93 ± 0.06^{aC}	13.53 ± 0.10^{aB}	13.79 ± 0.25^{aAB}	13.72 ± 0.14^{aAB}	13.78 ± 0.13^{abcA}
F4	12.69 ± 0.09^{cC}	13.44 ± 0.16^{abB}	13.54 ± 0.11^{aAB}	13.51 ± 0.19^{abcAB}	13.95 ± 0.17^{aA}
F5	12.89 ± 0.16^{abC}	13.50 ± 0.06^{abB}	13.72 ± 0.12^{aAB}	13.59 ± 0.11^{abcAB}	13.92 ± 0.18^{abA}

* Values represent the average $\pm$ standard deviation; All treatments, except FC1 (20% pork fat), have 7.2% of pork fat and 10.8% of linseed oil. [a, b, c, d, e] Means in the same column with the same lowercase letters did not differ significantly at P < 0.05. [A, B, C] Means in the same row with the same capital letters did not differ significantly at P < 0.05. FC1: Control (20% pork fat); FC2: Control with linseed oil. R: Rosemary; JP: Jamaica Pepper; WP: White Pepper; C: Coriander; M: Marjoram; S: Sage; T: Thyme. F1: 0.5% of blend 1 (R, JP, WP and C); F2: 0.5% of blend 2 (R, JP, WP, C and M); F3: 0.5% of blend 3 (R, JP, WP, C and S); F4: 0.5% of blend 4 (R, JP, WP, C and T); F5: 0.5% of blend 5 (R, JP, WP, C, M, S and T).

As shown in Table 4, the TBARS values of the bologna sausages made with the addition of herb and spice blends increased up to day 30 of storage, with a reduction after 45 days of storage.

52 A. K. Ferreira Ignácio Câmara and M. A. Rodrigues Pollonio

Table 4. TBARS values (mg malonaldehyde/kg sample) of the bologna sausages with different levels of linseed oil, pork fat, herbs and spices during storage

	0 day	15 days	30 days	45 days	60 days
TBARS					
FC1	0.099 ± 0.002^{cC}	0.076 ± 0.012^{cAB}	0.123 ± 0.010^{cA}	0.126 ± 0.007^{aAB}	0.097 ± 0.013^{aBC}
FC2	0.155 ± 0.003^{aC}	0.153 ± 0.039^{aAB}	0.139 ± 0.007^{bcA}	0.135 ± 0.022^{aAB}	0.121 ± 0.005^{aBC}
F1	0.084 ± 0.005^{cC}	0.110 ± 0.016^{bcAB}	0.196 ± 0.022^{aA}	0.138 ± 0.008^{aAB}	0.116 ± 0.025^{aBC}
F2	0.104 ± 0.009^{bcC}	0.128 ± 0.014^{abAB}	0.173 ± 0.006^{abA}	0.139 ± 0.014^{aAB}	0.122 ± 0.006^{aBC}
F3	0.090 ± 0.016^{cC}	0.130 ± 0.021^{abAB}	0.171 ± 0.017^{abA}	0.148 ± 0.015^{aAB}	0.125 ± 0.004^{aBC}
F4	0.104 ± 0.007^{bcC}	0.113 ± 0.010^{bcAB}	0.156 ± 0.023^{abcA}	0.130 ± 0.006^{aAB}	0.127 ± 0.015^{aBC}
F5	0.122 ± 0.004^{bC}	0.111 ± 0.014^{bcAB}	0.161 ± 0.012^{abcA}	0.131 ± 0.005^{aAB}	0.127 ± 0.008^{aBC}

* Values represent the average ± standard deviation; All treatments, except FC1 (20% pork fat), have 7.2% of pork fat and 10.8% of linseed oil. [a, b, c] Means in the same column with the same lowercase letters did not differ significantly at P < 0.05. [A, B, C] Means in the same row with the same capital letters did not differ significantly at P < 0.05. FC1: Control (20% pork fat); FC2: Control with linseed oil. R: Rosemary; JP: Jamaica Pepper; WP: White Pepper; C: Coriander; M: Marjoram; S: Sage; T: Thyme. F1: 0.5% of blend 1 (R, JP, WP and C); F2: 0.5% of blend 2 (R, JP, WP, C and M); F3: 0.5% of blend 3 (R, JP, WP, C and S); F4: 0.5% of blend 4 (R, JP, WP, C and T); F5: 0.5% of blend 5 (R, JP, WP, C, M, S and T).

The treatment FC2 containing linseed oil and without the addition of herbs and spices presented the highest TBARS value at day 15 of storage, followed by a reduction after this period. In contrast, the treatment FC1, with the addition of pork fat, exhibited an increase in TBARS values up to the day 45 of storage. This trend was also observed by Kim *et al.* (2016), who evaluated the oxidative stability of cooked chicken stored at 4 °C for 6 days and found an increase in TBARS values only up to day 4 of storage. According to Del Rio *et al.* (2005), the TBARS values decrease during storage, once several intermediate compounds of lipid oxidation, including MDA, possibly decompose into other oxidized compounds that do not react with TBA. Thus, in a shorter time, the sample FC2 had already

The Effects of Herbs and Spices on the Sensory ... 53

reached its highest oxidation, and the reduction of MDA at day 30 indicates that it had been consumed in other reactions.

Table 5. Texture profile analysis of the bologna sausages made with different levels of linseed oil, pork fat, herbs, and spices during storage

	0 day	15 days	30 days	45 days	60 days
Hardness (N)					
FC1	14.11 ± 0.69^{aC}	15.64 ± 0.71^{aB}	15.73 ± 0.60^{aB}	15.48 ± 0.23^{aB}	24.62 ± 0.82^{aA}
FC2	12.30 ± 0.19^{bC}	14.47 ± 0.55^{abB}	14.67 ± 0.42^{abB}	14.56 ± 0.87^{abB}	22.53 ± 0.85^{bA}
F1	12.65 ± 0.80^{bC}	14.56 ± 0.48^{abB}	15.13 ± 0.13^{abB}	14.94 ± 0.58^{abB}	22.77 ± 0.69^{bA}
F2	12.55 ± 0.23^{bC}	14.11 ± 0.35^{bB}	15.10 ± 0.09^{abB}	15.00 ± 0.37^{abB}	22.92 ± 0.27^{bA}
F3	12.38 ± 0.18^{bC}	14.29 ± 0.46^{bB}	14.93 ± 0.63^{abB}	14.72 ± 0.41^{abB}	22.28 ± 0.66^{bcA}
F4	12.42 ± 0.42^{bC}	14.10 ± 0.47^{bB}	14.39 ± 0.35^{bb}	14.47 ± 0.12^{abB}	22.42 ± 0.81^{bcA}
F5	12.34 ± 0.50^{bC}	14.22 ± 0.44^{bB}	14.63 ± 0.69^{bB}	14.14 ± 0.58^{bB}	20.84 ± 0.69^{cA}
Cohesiveness					
FC1	0.795 ± 0.00^{bAB}	0.792 ± 0.00^{dBC}	0.797 ± 0.01^{cC}	0.798 ± 0.01^{dABC}	0.799 ± 0.10^{bA}
FC2	0.816 ± 0.00^{aAB}	0.818 ± 0.01^{aBC}	0.812 ± 0.00^{aC}	0.822 ± 0.00^{aABC}	0.819 ± 0.30^{aA}
F1	0.816 ± 0.01^{aAB}	0.807 ± 0.01^{cBC}	0.802 ± 0.00^{bcC}	0.807 ± 0.00^{cABC}	0.816 ± 0.06^{aA}
F2	0.814 ± 0.02^{aAB}	0.811 ± 0.01^{bcBC}	0.807 ± 0.01^{abc}	0.814 ± 0.01^{bcABC}	0.816 ± 0.31^{aA}
F3	0.816 ± 0.00^{aAB}	0.814 ± 0.00^{abBC}	0.813 ± 0.00^{aC}	0.812 ± 0.01^{bcABC}	0.818 ± 0.13^{aA}
F4	0.814 ± 0.00^{aAB}	0.815 ± 0.01^{abBC}	0.813 ± 0.00^{aC}	0.817 ± 0.01^{abABC}	0.819 ± 0.17^{aA}
F5	0.819 ± 0.01^{aAB}	0.813 ± 0.00^{abBC}	0.814 ± 0.01^{aC}	0.815 ± 0.00^{bABC}	0.816 ± 0.18^{aA}
Chewiness (N)					
FC1	10.09 ± 0.62^{aD}	11.52 ± 0.23^{aC}	11.55 ± 0.31^{aBC}	11.73 ± 0.26^{aB}	17.63 ± 0.75^{aA}
FC2	9.24 ± 0.06^{bD}	10.99 ± 0.43^{abC}	10.93 ± 0.10^{aBC}	11.05 ± 0.74^{aB}	16.68 ± 0.54^{abA}
F1	9.79 ± 0.21^{abD}	11.21 ± 0.18^{abC}	11.20 ± 0.34^{aBC}	11.37 ± 0.52^{aB}	16.91 ± 0.62^{abA}
F2	9.55 ± 0.14^{abD}	10.59 ± 0.25^{bC}	11.26 ± 0.33^{aBC}	11.45 ± 0.39^{aB}	16.73 ± 0.80^{abA}
F3	9.33 ± 0.29^{bD}	10.73 ± 0.30^{abC}	10.97 ± 0.17^{aBC}	11.36 ± 0.36^{aB}	16.73 ± 0.82^{abA}
F4	9.57 ± 0.28^{abD}	10.76 ± 0.53^{abC}	10.82 ± 0.24^{aBC}	11.27 ± 0.16^{aB}	16.79 ± 0.24^{abA}
F5	9.27 ± 0.43^{bD}	10.97 ± 0.36^{abC}	11.01 ± 0.51^{aBC}	11.06 ± 0.40^{aB}	16.22 ± 0.39^{bA}

* Values represent the average ± standard deviation; All treatments, except FC1 (20% pork fat), have 7.2% of pork fat and 10.8% of linseed oil. [a, b, c, d] Means in the same column with the same lowercase letters did not differ significantly at P < 0.05. [A, B, C,D] Means in the same row with the same capital letters did not differ significantly at P < 0.05. FC1: Control (20% pork fat); FC2: Control with linseed oil. R: Rosemary; JP: Jamaica Pepper; WP: White Pepper; C: Coriander; M: Marjoram; S: Sage; T: Thyme. F1: 0.5% of blend 1 (R, JP, WP and C); F2: 0.5% of blend 2 (R, JP, WP, C and M); F3: 0.5% of blend 3 (R, JP, WP, C and S); F4: 0.5% of blend 4 (R, JP, WP, C and T); F5: 0.5% of blend 5 (R, JP, WP, C, M, S and T).

Among the treatments with herb and spice blends, the best antioxidant protection was observed in F4 (rosemary, Jamaica, white pepper, coriander, and thyme) and F5 (same herbs of F4 plus marjoram and sage), with no significant differences (P <0.05) when compared to the control

FC1 (pork fat) throughout the storage period. When comparing the treatments F4 and F5 with the other treatments, it is observed the presence of thyme probably led to a better antioxidant protection. Thyme (*Thymus vulgaris* L.) is an aromatic herb of the family *Lamiaceae*, which contains carvacrol and thymol as main phenolic compounds with proven antioxidant effects (Dogu-Baykut *et al.*, 2014; Tohidi *et al.*, 2017).

The results of firmness, cohesiveness, and chewiness of the samples are shown in Table 5. In general, the treatments with addition of linseed oil had lower firmness and chewiness. The effects of the addition of vegetable oils in emulsified meat products have led to conflicting results, being quite complex since it will depend on several factors, including the level of fat replacement by oil, the protein content of the formulation, the moisture content, and the form of incorporation of oil into the meat batter (directly, pre-emulsified, or encapsulated, for example). Some authors, such as Lurueña-Martínez *et al.* (2004), observed that the addition of olive oil to low-fat frankfurters led to a decrease in firmness and chewiness of the products, which was also observed in the present study. In contrast, Delgado-Pando *et al.* (2010) found higher firmness and chewiness in frankfurters with fat partially replaced by a combination of linseed, fish, and olive oils stabilized in different protein systems. Câmara and Pollonio (2015) evaluated the effects of replacing pork fat (5 to 20%) by linseed oil (2.5 to 12.5%) on the texture properties of bologna sausages using a central rotational compound design, and found higher firmness in the treatment containing high total fat (20% pork fat and 7.5% linseed oil), with no significant differences (P <0.05) from the control (20% pork fat).

In relation to the storage time, similar behavior was observed for the parameters firmness and chewiness, which increased slightly during the storage, with a significant increase at day 60. López-López *et al.* (2009) studied frankfurters with 50% replacement of pork fat by olive oil, and also observed a significant increase in firmness and chewiness after 41 days of chilled storage. As reported by Andrés *et al.* (2006), the increase in firmness during the storage may be due to purge losses. Concerning the cohesiveness of the samples, the treatment FC1 was the least cohesive throughout the storage period, differing from the other treatments. Similar

The Effects of Herbs and Spices on the Sensory ... 55

results were also found by Cáceres *et al.* (2008), who observed greater cohesiveness in sausages containing oils pre-emulsified with sodium caseinate, probably due to the strong interaction between these proteins, which are excellent emulsifiers, with the other emulsion constituents.

3.3. Fatty Acid Analysis

According to the results in Table 6, MUFAs (monounsaturated fatty acids) were the most abundant fatty acids in the treatment containing pork fat (FC1), followed by SFAs (saturated fatty acids). The addition of linseed oil to the bologna sausages formulations led to important changes in the lipid profiles of the products, mainly due to the composition of this oil, rich in PUFAs, especially ω-3 fatty acids. Significant reductions of SFAs were observed for all treatments when compared with the control (FC1). In addition, the total PUFAs increased from 13.90% in FC1 to about 50% in the other treatments with linseed oil, corresponding to an increase of more than 70%. Among the treatments with the addition of linseed oil, some herb and spice blends stood out by maintaining the PUFAs levels. The highest levels of α-linolenic acid (C18: 3) of the treatments with modified lipid profile were found in F2 and F4, with 36.16% and 35.85% respectively, while FC2 presented the lowest level, with 34.62%, with significant differences (P <0.05) when compared to the other treatments.

Ansorena & Astiasarán (2004) studied fermented sausages with 25% fat replaced by linseed oil, and found high linoleic and α-linolenic acids levels in the products containing the synthetic antioxidants BHA and BHT, indicating that the antioxidants can decrease the oxidative deterioration of fatty acids, which are more susceptible to oxidation.

As the intake of saturated fats has led to health implications, studies have recommended specific fatty acids ratios, and the recommended PUFA/SFA ratio should be greater than 0.4, while the ω-6 / ω-3 ratio should be equal to or less than 5: 1 (Simopoulos, 2004; Bernardi *et al.*, 2016).

Table 6. Fatty acid composition (%) of the bologna sausages made with different levels of linseed oil, pork fat, herbs, and spices at the beginning of storage

Fatty acid	Treatments						
	FC1	FC2	F1	F2	F3	F4	F5
Myristic C14:0	1.36 ± 0.00[a]	0.64 ± 0.01[b]	0.58 ± 0.00[d]	0.58 ± 0.00[d]	0.61 ± 0.00[c]	0.61 ± 0.00[c]	0.63 ± 0.00[bc]
Palmitic C16:0	23.55 ± 0.00[a]	12.96 ± 0.07[b]	12.54 ± 0.01[c]	12.05 ± 0.09[d]	12.25 ± 0.07[d]	12.21 ± 0.04[d]	12.86 ± 0.02[b]
Stearic C18:0	13.45 ± 0.05[a]	7.33 ± 0.05[b]	7.04 ± 0.02[c]	6.67 ± 0.08[d]	6.79 ± 0.05[d]	6.80 ± 0.02[d]	7.25 ± 0.02[b]
Arachidic C20:0	0.25 ± 0.00[a]	0.16 ± 0.00[b]	0.15 ± 0.00[bc]	0.14 ± 0.00[d]	0.14 ± 0.00[cd]	0.14 ± 0.00[cd]	0.16 ± 0.00[b]
Others SFA's	0.86 ± 0.01[a]	0.57 ± 0.03[b]	0.56 ± 0.00[b]	0.62 ± 0.00[b]	0.63 ± 0.01[b]	0.61 ± 0.00[b]	0.59 ± 0.02[b]
Σ SFA	**39.47**	**21.66**	**20.87**	**20.06**	**20.42**	**20.37**	**21.49**
Palmitoleic C16:1	2.22 ± 0.01[a]	1.02 ± 0.00[c]	1.0 ± 0.00[d]	1.01 ± 0.00[cd]	1.04 ± 0.00[b]	1.02 ± 0.00[c]	1.01 ± 0.00[cd]
Margaroleic C17:1	0.45 ± 0.00[a]	0.23 ± 0.00[bc]	0.22 ± 0.00[c]	0.25 ± 0.01[b]	0.24 ± 0.00[bc]	0.23 ± 0.00[bc]	0.23 ± 0.00[bc]
Oleic C18:1	42.84 ± 0.01[a]	27.61 ± 0.10[b]	27.36 ± 0.06[c]	27.12 ± 0.01[d]	27.41 ± 0.09[c]	27.14 ± 0.04[d]	27.25 ± 0.01[cd]
Eicosenoic C20:1	0.80 ± 0.00[a]	0.40 ± 0.00[b]	0.38 ± 0.00[c]	0.33 ± 0.00[e]	0.34 ± 0.00[de]	0.35 ± 0.01[d]	0.38 ± 0.00[c]
Σ MUFA	**46.31**	**29.26**	**28.96**	**28.71**	**29.03**	**28.74**	**28.87**
Linoleic C18:2 (n-6)	12.63 ± 0.04[d]	13.70 ± 0.11[bc]	13.97 ± 0.00[b]	14.42 ± 0.09[a]	14.40 ± 0.04[a]	14.38 ± 0.08[a]	13.65 ± 0.05[c]
Linolenic C18:3 (n-3)	0.61 ± 0.01[e]	34.62 ± 0.03[d]	35.57 ± 0.07[bc]	36.16 ± 0.08[a]	35.33 ± 0.22[c]	35.86 ± 0.03[ab]	35.23 ± 0.10[c]
Arachidonic C20:4 (n-6)	0.36 ± 0.00[a]	0.24 ± 0.00[b]	0.24 ± 0.00[b]	0.26 ± 0.01[b]	0.24 ± 0.00[b]	0.25 ± 0.00[b]	0.24 ± 0.01[b]
Stearidonic C18:4 (n-3)	0.14 ± 0.01[a]	0.10 ± 0.01[c]	0.10 ± 0.00[c]	0.11 ± 0.00[bc]	0.12 ± 0.01[ab]	0.11 ± 0.00[bc]	0.09 ± 0.00[c]
EPA C20:5 (n-3)	0.05 ± 0.02[a]	0.04 ± 0.01[a]	0.04 ± 0.01[a]	0.03 ± 0.00[a]	0.03 ± 0.00[a]	0.03 ± 0.00[a]	0.04 ± 0.00[a]
DPA C22:5 (n-3)	0.11 ± 0.02[a]	0.10 ± 0.02[a]	0.09 ± 0.01[a]	0.09 ± 0.01[a]	0.07 ± 0.00[a]	0.08 ± 0.02[a]	0.10 ± 0.02[a]
Σ PUFA	**13.90**	**48.80**	**50.01**	**51.07**	**50.19**	**50.71**	**49.35**
PUFA/SFA	**0.35**	**2.25**	**2.40**	**2.55**	**2.46**	**2.49**	**2.30**
Σ n-3	**0.91**	**34.86**	**35.80**	**36.39**	**35.55**	**36.08**	**35.46**
Σ n-6	**12.99**	**13.94**	**14.21**	**14.68**	**14.64**	**14.63**	**13.89**
n-6/n-3	**14.27**	**0.39**	**0.40**	**0.40**	**0.41**	**0.41**	**0.39**
Elaidic C18:1t	0.20	0.11	-	-	0.18	-	0.13
Linolenic C18:3t	0.12 ± 0.00[b]	0.17 ± 0.01[a]	0.16 ± 0.00[ab]	0.16 ± 0.01[ab]	0.18 ± 0.02[a]	0.18 ± 0.00[a]	0.16 ± 0.00[ab]
Σ Trans	**0.32**	**0.28**	**0.16**	**0.16**	**0.36**	**0.18**	**0.29**

* Values represent the average ± standard deviation; All treatments, except FC1 (20% pork fat), have 7.2% of pork fat and 10.8% of linseed oil. [a,b,c,d,e] Means in the same column with the same letters did not differ significantly at $P < 0.05$. FC1: Control (20% pork fat); FC2: Control with linseed oil. R: Rosemary; JP: Jamaica Pepper; WP: White Pepper; C: Coriander; M: Marjoram; S: Sage; T: Thyme. F1: 0.5% of blend 1 (R, JP, WP and C); F2: 0.5% of blend 2 (R, JP, WP, C and M); F3: 0.5% of blend 3 (R, JP, WP, C and S); F4: 0.5% of blend 4 (R, JP, WP, C and T); F5: 0.5% of blend 5 (R, JP, WP, C, M, S and T).

Table 7. Fatty acid composition (%) of the bologna sausages made with different levels of linseed oil, pork fat, herbs, and spices at the end of storage (60 days)

Fatty acid	Treatments						
	FC1	FC2	F1	F2	F3	F4	F5
Myristic C14:0	1.35 ± 0.00[a]	0.65 ± 0.00[c]	0.61 ± 0.01[d]	0.63 ± 0.01[cd]	0.71 ± 0.00[b]	0.62 ± 0.01[cd]	0.70 ± 0.01[b]
Palmitic C16:0	23.69 ± 0.23[a]	12.92 ± 0.04[c]	13.06 ± 0.12[c]	12.71 ± 0.24[c]	13.91 ± 0.09[b]	12.45 ± 0.16[c]	14.02 ± 0.16[b]
Stearic C18:0	13.50 ± 0.04[a]	7.77 ± 0.02[bc]	7.49 ± 0.09[cd]	7.24 ± 0.15[de]	7.91 ± 0.15[bc]	7.01 ± 0.09[e]	8.09 ± 0.08[b]
Arachidic C20:0	0.26 ± 0.00[a]	0.16 ± 0.00[bc]	0.16 ± 0.00[bc]	0.14 ± 0.00[d]	0.15 ± 0.00[cd]	0.14 ± 0.00[d]	0.17 ± 0.00[b]
Others SFA's	0.93 ± 0.00[a]	0.67 ± 0.00[cd]	0.63 ± 0.00[d]	0.76 ± 0.00[bc]	0.94 ± 0.04[a]	0.68 ± 0.03[cd]	0.79 ± 0.03[b]
Σ SFA	39.73	22.17	21.95	21.48	23.62	20.90	23.77
Palmitoleic C16:1	2.20 ± 0.01[a]	1.14 ± 0.00[bc]	1.06 ± 0.02[d]	1.06 ± 0.01[d]	1.15 ± 0.00[b]	1.04 ± 0.01[d]	1.09 ± 0.00[cd]
Margaroleic C17:1	0.45 ± 0.00[a]	0.23 ± 0.00[cd]	0.22 ± 0.00[d]	0.24 ± 0.00[c]	0.26 ± 0.00[b]	0.24 ± 0.00[c]	0.24 ± 0.00[c]
Oleic C18:1	42.62 ± 0.13[a]	27.31 ± 0.05[d]	28.14 ± 0.03[cd]	28.14 ± 0.23[cd]	29.69 ± 0.01[b]	27.47 ± 0.30[d]	28.71 ± 0.26[c]
Eicosenoic C20:1	0.81 ± 0.00[a]	0.44 ± 0.00[b]	0.41 ± 0.00[b]	0.38 ± 0.01[c]	0.41 ± 0.01[b]	0.37 ± 0.01[c]	0.44 ± 0.00[b]
Σ MUFA	46.08	29.12	29.83	29.82	31.51	29.12	30.48
Linoleic C18:2 (n-6)	12.23 ± 0.07[d]	14.05 ± 0.02[b]	13.96 ± 0.04[b]	14.60 ± 0.01[a]	14.01 ± 0.03[b]	14.57 ± 0.09[a]	13.47 ± 0.08[c]
Linolenic C18:3 (n-3)	0.60 ± 0.01[d]	32.86 ± 0.11[ab]	33.28 ± 0.25[ab]	33.04 ± 0.00[ab]	29.49 ± 0.32[c]	34.47 ± 0.36[a]	31.04 ± 0.40[bc]
Arachidonic C20:4 (n-6)	0.28 ± 0.00[a]	0.21 ± 0.00[b]	0.20 ± 0.00[bc]	0.20 ± 0.01[bc]	0.17 ± 0.01[c]	0.22 ± 0.01[b]	0.18 ± 0.00[bc]
Stearidonic C18:4 (n-3)	0.50 ± 0.01[b]	0.87 ± 0.01[a]	0.38 ± 0.02[b]	0.46 ± 0.02[b]	0.54 ± 0.22[ab]	0.35 ± 0.00[b]	0.46 ± 0.00[b]
EPA C20:5 (n-3)	0.06 ± 0.00[a]	0.09 ± 0.01[a]	0.06 ± 0.01[a]	0.06 ± 0.01[a]	0.06 ± 0.01[a]	0.06 ± 0.02[a]	0.07 ± 0.00[a]
DPA C22:5 (n-3)	0.11 ± 0.00[a]	0.12 ± 0.01[a]	0.09 ± 0.01[a]	0.08 ± 0.01[a]	0.11 ± 0.02[a]	0.07 ± 0.01[a]	0.10 ± 0.01[a]
DHA C22:6 (n-3)	0.11 ± 0.00[a]	0.22 ± 0.00[a]	0.09 ± 0.00[a]	0.11 ± 0.01[a]	0.19 ± 0.11[a]	0.08 ± 0.00[a]	0.13 ± 0.01[a]
Σ PUFA	13.89	48.42	48.06	48.55	44.57	49.82	45.45
PUFA/SFA	0.35	2.18	2.19	2.26	1.88	2.38	1.91
Σ n-3	1.38	34.16	33.90	33.75	30.39	35.03	31.80
Σ n-6	12.51	14.26	14.16	14.80	14.18	14.79	13.65
n-6/n-3	9.06	0.42	0.42	0.44	0.47	0.42	0.43
Elaidic C18:1t	0.18	0.13	-	-	0.14	-	0.11
Linolenic C18:3t	0.12 ± 0.00[c]	0.16 ± 0.00[b]	0.16 ± 0.02[b]	0.15 ± 0.00[b]	0.16 ± 0.01[ab]	0.16 ± 0.00[b]	0.19 ± 0.02[a]
Σ Trans	0.30	0.29	0.16	0.15	0.30	0.16	0.30

* Values represent the average ± standard deviation; All treatments, except FC1 (20% pork fat), have 7.2% of pork fat and 10.8% of linseed oil. [a,b,c,d,e] Means in the same column with the same letters did not differ significantly at $P < 0.05$. FC1: Control (20% pork fat); FC2: Control with linseed oil. R: Rosemary; JP: Jamaica Pepper; WP: White Pepper; C: Coriander; M: Marjoram; S: Sage; T: Thyme. F1: 0.5% of blend 1 (R, JP, WP and C); F2: 0.5% of blend 2 (R, JP, WP, C and M); F3: 0.5% of blend 3 (R, JP, WP, C and S); F4: 0.5% of blend 4 (R, JP, WP, C and T); F5: 0.5% of blend 5 (R, JP, WP, C, M, S and T).

58   *A. K. Ferreira Ignácio Câmara and M. A. Rodrigues Pollonio*

All treatments with the addition of linseed oil are in accordance with the recommended standards, with PUFA/SFA ratio ranging from 2.25 (FC2) to 2.55 (F2), and ω-6/ω-3 ratio values around 0.4, opposite to the value found in the control treatment (FC1), with 14.27. Selani *et al.* (2016) reported that the addition of canola oil modified the fatty acid profile of hamburgers, increasing the PUFA/SFA ratio and decreasing ω-6/ω-3 ratio, thus providing a better nutritional quality to the meat products.

Table 7 shows the fatty acid profile of bologna sausages after 60 days of refrigerated storage, which allowed the evaluation of the nutritional quality of these products at the end of the shelf life. After 2 months of storage, high PUFAs levels remained in the treatments with addition of linseed oil. The treatments F4 and F2 had total PUFAs slightly higher than FC2 (48.42%), with 49.82% and 48.55%, respectively. Valencia *et al.* (2006) evaluated *Chorizo de Pamplona* with 25% substitution of pork fat by linseed oil and synthetic antioxidants (BHA and BHT) and concluded that the products had their nutritional benefits preserved during 5 months of storage.

3.4. Sensory Evaluation

The sensory evaluation indicated that the herb and spice blends positively affected the sensory attributes of the bologna sausages with modified lipid profiles (Figure 3). The results of the overall impression indicated good consumer's acceptance of the treatments with the addition herbs and spices, with no significant differences (P < 0.05) when compared to the control sample with the addition of pork fat (FC1). Jiménez-Colmenero *et al.* (2010) also found similar acceptance ratings for the control treatment when compared to the samples containing olive oil pre-emulsions stabilized with different proteins.

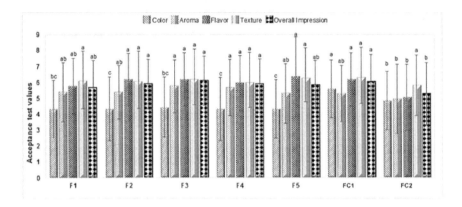

Figure 3. Acceptance scores of the bologna sausages made with the addition of different levels of linseed oil, pork fat, herbs, and spices. Values represent the average ± standard deviation; All treatments have 7.2% pork fat and 10.8% linseed oil, except FC1 (20% pork fat). [a,b,c] Means in the same sensory attribute with the same letters did not differ significantly at P<0.05 (Tukey's test). F1: 0.5% of blend 1 (R, JP, WP, and C); F2: 0.5% of blend 2 (R, JP, WP, C, and M); F3: 0.5% of blend 3 (R, JP, WP, C, and S); F4: 0.5% of blend 4 (R, JP, WP, C, and T); F5: 0.5% of blend 5 (R, JP, WP, C, M, S, and T); FC1: Control (20% pork fat); FC2: Control with linseed oil. R: Rosemary; JP: Jamaica Pepper; WP: White Pepper; C: Coriander; M: Marjoram; S: Sage; T: Thyme.

Lower acceptance scores were observed for the attribute color, evidencing that the herb and spice blends interfered in this attribute. The lowest score (4.33) was observed in the sample F4, which did not differ (P <0.05) from the formulations F1, F2, F3, and F5, while the control sample FC1 exhibited the highest score (5.61). These results indicate that regardless of the blend composition, the assessors found similar color interferences. For the treatments with the addition of herb and spice blends, higher scores were observed for the attributes aroma and flavor when compared to the FC2 containing linseed oil, with no significant differences (P <0.05) from the control formulation made with the addition of pork fat (FC1). Chinprahast *et al.* (2012) also found that the addition of rosemary leaves (3.2%) improved the flavor and oxidative stability of pork patties. Valencia *et al.* (2008) investigated the sensory properties of pork sausages with fat substitution (15%) by linseed oil or fish oil and the addition of green tea catechins and green coffee as antioxidants. According to those authors, the sensory properties of the sausages were not affected by the

60 *A. K. Ferreira Ignácio Câmara and M. A. Rodrigues Pollonio*

addition of linseed oil, and the green tea catechins improved the acceptability of the sausages made with the addition of fish oil. Regarding the attribute texture, the assessors did not identify differences between treatments.

CONCLUSION

Bologna sausages made with the replacement of 54% pork fat by linseed oil, and addition of herb and spice blends were stable to lipid oxidation during the refrigerated storage. Among the treatments containing herb and spice blends, a better antioxidant protection was observed in F4 (rosemary, Jamaica, white pepper, coriander, and thyme) and F5 (same herbs of F4 plus marjoram and sage), highlighting the presence of thyme, which may have contributed to a better antioxidant protection of the bologna sausage produced with the addition of linseed oil. In the sensory evaluation, all herbs and spices of this study provided higher scores for the attributes flavor, aroma, texture, and overall impression, with no difference between the treatments and the control formulation made with pork fat, except for the attribute color, which was negatively affected by the presence of linseed oil and herbs and spices.

The manufacture of bologna sausages with better lipid profile containing herb and spice blends has proven to be viable under the technological and sensory point of view, and nutritionally superior to the conventional meat products, since it allowed an increase of more than 70% in the PUFAs levels and a reduction of approximately 47% in the SFAs levels, when compared with the control sample.

REFERENCES

Aggarwal, M., Aggarwal, B. & Rao, J. (2017). Integrative medicine for cardiovascular disease and prevention. *The Medical Clinics of North America, 101*, 895-923.

The Effects of Herbs and Spices on the Sensory ... 61

Alejandre, M., Passarini, D., Astiasarán, I. & Ansorena, D. (2017). The effect of low-fat beef patties formulated with a low-energy fat analogue enriched in long-chain polyunsaturated fatty acids on lipid oxidation and sensory attributes. *Meat Science*, *134*, 7-13.

Andrés, S.C., García, M.E., Zaritzky, N.E., & Califano, A.N. (2006). Storage stability of low-fat chicken sausages. *Journal of Food Engineering*, *72*(4), 311-319.

Ansorena, D., & Astiasarán, I. (2004). The use of linseed oil improves nutritional quality of the lipid fraction of dry-fermented sausages. *Food Chemistry*, *87*(1), 69–74.

AOAC (2005). *Official methods of analysis of AOAC International* (18th ed.). Maryland, USA: Association of Official Analytical Chemistry.

AOCS (2004). *Official Methods and Recommended Practices of the American Oil Chemists' Society* (5th ed.). Champaign: American Oil Chemists' Society.

Bernardi, D.M., Bertol, T.M., Pflanzer, S.B., Sgarbieri, V.C., & Pollonio, M.A.R. (2016). ω-3 in meat products: benefits and effects on lipid oxidative stability. *Journal of the Science of Food and Agriculture*, *96*(8), 2620-2634.

Bligh, E.G., & Dyer, W.J. (1959). A rapid method of total lipid extraction and purification. *Canadian Journal Biochemical Physiology, 37*, 911-914.

Bohrer, B.M. (2017). Review: Nutrient density and nutritional value of meat products and non-meat foods high in protein. *Trends in Food Science & Technology*, 65, 103-112.

Brasil (2000). Instrução Normativa n° 04, de 05 de abril de 2000. *Regulamento Técnico de Identidade e Qualidade de Mortadela.* [Normative Instruction n° 04, of April 5, 2000. Technical Regulation of Identity and Quality of Mortadella]. Department of Agriculture and Food Supply (Available at: http://www.agricultura.gov.br/sislegis. Acess:13/07/2016).

Cáceres, E., García, M.L., Selgas, M.D. (2008). Effect of pre-emulsified fish oil – as source of PUFA n-3 – on microstructure and sensory properties of mortadella, a Spanish bologna-type sausage. *Meat Science, 80*(2),183-193.

Câmara, A.K.F.I, & Pollonio, M.A.R. (2015). Reducing animal fat in bologna sausage using pre-emulsified linseed oil: technological and sensory properties. *Journal of Food Quality, 38*(3), 201-212.

Chinprahast, N., Suwannadath, A. & Homjabok, T. (2012). Use of rosemary (*Rosmarinus officinalis* L.) leaf for improving oxidative stability of microwave-precooked traditional Thai pork patty and its frozen storage trial. *International Journal of Food Science and Technology*, 47, 2165-2174.

Del Rio, D., Stewart, A.J., & Pellegrini, N. (2005). A review of recent studies on malondialdehyde as toxic molecule and biological marker of oxidative stress. *Nutrition, Metabolism and Cardiovascular Diseases, 15*(4), 316-328.

Delgado-Pando, G., Cofrades, S., Ruiz-Capillas, C., & Jiménez-Colmenero, F. (2010). Healthier lipid combination as functional ingredient influencing sensory and technological properties of low-fat frankfurters. *European Journal of Lipid Science and Technology, 112*(8), 859-870.

Dogu-Baykut, E., Gunes, G. & Decker, E.A. (2014). Impact of shortwave ultraviolet (UV-C) radiation on the antioxidante activity of thyme (*Thymus vulgaris* L.). *Food Chemistry,* 157, 167-173.

Dolatabadi, J.E.N. & Kashanian, S. (2010). A review on DNA interaction with synthetic phenolic food additives. *Food Research International*, 43, 1223-1230.

Erdmann, M.E., Lautenschlaeger, R., Zeeb, B., Gibis, M. & Weiss, J. (2017). Effect of differently sized O/W emulsions loaded with rosemary extract on lipid oxidation in cooked emulsion-type sausages rich in n-3 fatty acids. *LWT - Food Science and Technology,* 79, 496-502.

Exarchou, V., Nenadis, N., Tsimidou, M., Gerothanassis, I.P., Troganis, A., & Boskou, D. (2002). Antioxidant activities and phenolic composition of extracts from greek oregano, greek sage, and summer savory. *Journal of Agricultural and Food Chemistry, 50*(19), 5294-5299.

Gallardo, B., Manca, M.G., Mantecón, A.R., Nudda, A., & Manso, T. (2015). Effects of linseed oil and natural or synthetic vitamin E supplementation in lactating ewes' diets on meat fatty acid profile and lipid oxidation from their milk fed lambs. *Meat Science, 102*, 79–89.

Grasso, S., Brunton, N.P., Lyng, J.G., Lalor, F. & Monahan, F.J. (2014). Healthy processed meat products - Regulatory, reformulation and consumer challenges. *Trends in Food Science & Technology,* 39, 4-17.

Gray, J.I., & Pearson, A.M. (1987). Rancidity and warmed-over flavor. In A.M. Pearson, & J.I. Gray (Eds.), *Advances in Meat Research* (pp. 221-269). New York: Van Nostrand Reinhold Company.

Hernández-Hernández, E., Ponce-Alquicira, E., Jaramillo Flores, M.E., & Legarreta, G.I. (2009). Antioxidant effect rosemary (*Rosmarinus officinalis L.*) and orégano (*Origanum vulgare L.*) extracts on TBARS and colour of model raw pork batters. *Meat Science, 81*(2), 410–417.

Houssain, M.B., Patras, A., Barry-Ryan, C., Martin-Diana, A.B, & Brunton, N.P. (2011). Application of principal component and hierarchical cluster analysis to classify different spices based on in vitro antioxidant activity and individual polyphenolic antioxidante compounds. *Journal of Functional Foods, 03*, 179-189.

Hughes, E., Cofrades, S., & Troy, D.J. (1997). Effects of fat level, oat fiber and carrageenan on frankfurters formulated with 5, 12 and 30% fat. *Meat Science, 45*(3), 273-281.

Jacobsen, C., Let, M.B., Nielsen, N.S., & Meyer, A.S. (2008). Antioxidant strategies for preventing oxidative flavour deterioration of foods enriched with n-3 polyunsaturated lipids: a comparative evaluation. *Trends in Food Science and Technology, 19*(2), 76-93.

Jiang, J., & Xiong, Y.L. (2016). Natural antioxidants as food and feed additives to promote health benefits and quality of meat products: A review. *Meat Science, 120*, 107-117.

Jiménez-Colmenero, F., Triki, M., Herrero, A., Rodríguez-Salas, L., Ruiz-Capillas, C. (2013). Healthy oil combination stabilized in a konjac matrix as pork fat replacement in low-fat, PUFA-enriched, dry

64   *A. K. Ferreira Ignácio Câmara and M. A. Rodrigues Pollonio*

fermented sausages. *LWT-Food Science and Technology, 51*(1), 158-163.

Jiménez-Colmenero, F., Herrero, A., Pintado, T., Solas, M.T., & Ruiz-Capillas, C. (2010). Influence of emulsified olive oil stabilizing system used for pork backfat replacement in frankfurters. *Food Research International, 43*(8), 2068-2076.

Jinap, S., Iqbal, S.Z., & Selvam, R.M.P. (2015). Effect of selected local spices marinades on the reduction of heterocyclic amines in grilled beef (satay). *LWT-Food Science and Technology, 63*(2), 919-926.

Jung, E., & Joo, N. (2013). Roselle (*Hibiscus sabdariffa L.*) and soybean oil effects on quality characteristics of pork patties studied by response surface methodology. *Meat Science, 94*(3), 391–401.

Karpinska, M., Borowski, J., & Danowska-Oziewicz, M. (2001). The use of natural antioxidants in ready-to-serve food. *Food Chemistry, 72*(1), 5-9.

Kim, S.Y., Li, J., Lim, N.R., Kang, B.S. & Park, H.J. (2016). Prediction of warmed-over flavour development in cooked chicken by colorimetric sensor array. *Food Chemistry, 211*, 440-447.

Kong, B., Zhang, H., & Xiong, Y.L. (2010). Antioxidant activity of spice extracts in a liposome system and in cooked pork patties and the possible mode of action. *Meat Science, 85*(4), 772-778.

Lebret, B., Lefaucheur, L., Mourot, J. & Bonneau, M. (1996). Influence des facteurs d 'élevage sur la qualité de la viande de porc. *Journées de la Recherche Porcine en France, 28*, 137-156. [Influence of breeding factors on the quality of pork meat. *Pork Research Days in France, 28*, 137-156].

López-López, l., Cofrades, S., & Jiménez-Colmenero, F. (2009). Low-fat frankfurters enriched with n-3 PUFA and edible seaweed: Effects of olive oil and chilled storage on physicochemical, sensory and microbial characteristics. *Meat Science, 83*(1), 148-154.

Lu, P., Zhang, L.Y., Yin, J.D., Everts, A.K.R. & Li, D.F. (2008). Effects of soybean oil and linseed oil on fatty acid compositions of muscle lipids and cooked pork flavor. *Meat Science, 80*, 910-918.

Lurueña-Martínez, M.A., Vivar-Quintana, A.M., & Revilla, I. (2004) Effect of locust bean/xanthan gum addition and replacement of pork fat with olive oil on the quality characteristics of low-fat frankfurters. *Meat Science, 68*(3), 383-389.

Negre-Salvayre, A., Coatrieux, C., Ingueneau, C., & Salvayre, R. (2008). Advanced lipid peroxidation end products in oxidative damage to proteins. Potential role in diseases and therapeutic prospects for the inhibitors. *British Journal of Pharmacology, 153*(1), 6–20.

Ninfali, P., Mea, G., Giorgini, S., Rocchi, M., & Bacchiocca, M. (2005). Antioxidant capacity of vegetables, spices and dressings relevant to nutrition. *British Journal of Nutrition, 93*(2), 257-266.

Paneras, E.D., Bloukas, J.G., & Filis, D.G. (1998). Production of low-fat frankfurters with vegetable oils following the dietary guidelines for fatty acids. *Journal of Muscle Foods, 9*(2), 111-126.

Przygodzka, M., Zielinska, D., Ciesarová, Z., Kukurová, K., & Zielinski, H. (2014). Comparison of methods for evaluation of the antioxidant capacity and phenolic compounds in common spices. *LWT-Food Science and Technology, 58*(2), 321-326.

Raes, K., De Smet, S., Demeyer, D. (2004). Effect of dietary fatty acids on incorporation of long chain polyunsaturated fatty acids and conjugated linoleic acid in lamb, beef and pork meat: A review. *Animal Feed Science and Technology*, 113,199-221.

Salcedo-Sandoval, L; Ruiz-Capillas, C; Cofrades, S.; Triki, M.; Jiménez-Colmenero, F. (2015). Shelf-life of n-3 PUFA enriched frankfurters formulated with a konjac-based oil bulking agent. *LWT-Food Science and Technology, 62*(1), 711-717.

Salih, A.M., Smith, D.M., Price, J.F., & Dawson, L.E. (1987). Modified extraction method for 2-thiobarbituric acid method for measuring lipid oxidation in poultry. *Poultry Science, 66*(9), 1483–1488.

SAS (2002). *SAS User's guide*. V.9.1. Cary, NC. USA: Statistical Analysis Systems Institute, Inc.

Schaefer, E.J. (2002). Lipoproteins, nutrition, and heart disease. *The American Journal of Clinical Nutrition, 75*(2), 191-212.

66   *A. K. Ferreira Ignácio Câmara and M. A. Rodrigues Pollonio*

Selani, M.M., Shirado, G.A.N., Margiotta, G.B., Rasera, M.L., Marabesi, A.C., Piedade, S.M.S., Contreras-Castillo, C.J., & Canniatti-Brazaca, S.G. (2016). Pineapple by-product and canola oil as partial fat replacers in low-fat beef burger: Effects on oxidative stability, cholesterol content and fatty acid profile. *Meat Science, 115*, 9-15.

Shan, B., Cai, Y.Z., Sun, M., & Corke, H. (2005). Antioxidant Capacity of 26 Spice Extracts and Characterization of Their Phenolic Constituents. *Journal of Agriculture and Food Chemistry, 53*(20), 7749-7759.

Simopoulos, A.P. (2004). Omega-6/Omega-3 Essential fatty acid ratio and chronic diseases. *Food Reviews International, 20(1)*, 77-90.

Slavin, M., Dong, M., Gewa, C. (2016). Effect of clove extract pretreatment and drying conditions on lipid oxidation and sensory discrimination of dried omena (*Rastrineobola argentea)* fish. *International Journal of Food Science & Technology,* 51, 2376-2385.

Stone, H.S., & Sidel, J.L. (1993). *Sensory evaluation practices* (2th ed.) London: Academic Press.

Tohidi, B., Rahimmalek, M. & Arzani, A. (2017). Essential oil composition, total phenolic, flavonoid contents, and antioxidant activity of Thymus species collected from different regions of Iran. *Food Chemistry,* 220, 153-161.

Trindade, R.A., Mancini-Filho, J., & Villavicencio, A.LC.H. (2010). Natural antioxidants protecting irradiated beef burgers from lipid oxidation. *LWT-Food Science and Technology, 43*(1), 98-104.

Uhl, S.R. (2000). *Handbook of Spices, Seasonings and Flavorings.* Florida: CRC Press LLC.

Valencia, I., O'Grady, M.N., Ansorena, D., Astiasarán, I., Kerry, J.P. (2008). Enhancement of the nutritional status and quality of fresh pork sausages following the addition of linseed oil, fish oil and natural antioxidants. *Meat Science,* 80, 1046-1054.

Valencia, I., Ansorena, D., & Astiasarán, I. (2006). Stability of linseed oil and antioxidants containing dry fermented sausages: A study of the lipid fraction during different storage conditions. *Meat Science, 73*(2), 269-277.

The Effects of Herbs and Spices on the Sensory ...   67

Valldervú-Queralt, A., Regueiro, J., Martinez-Huelano, M., Alvarenga, J.F.R., Leal, L.N., & Lamuela-Raventos, R.M. (2014). A comprehensive study on the phenolic profile of widely used culinary herbs and spices: Rosemary, thyme, oregano, cinnamon, cumin and bay. *Food Chemistry,* 154, 299-307.

Verhagen, H., Schilderman, P.E.L., Kleinjans, J.C.S. Butylated hydroxyanisole in perspective. (1991). *Chemico-Biological Interactions,* 80, 109-134.

Wangensteen, H., Samuelsen, A.B., & Malterud, K.E. (2004). Antioxidant activity in extracts from coriander. *Food Chemistry, 88*(2), 293-297.

WHO - World and Health Organization. (2003). Diet, nutrition and the prevention of chronic diseases, *WHO Technical Report Series,* 916, 160 p.

Wong, P.Y., & Kitts, D.D. (2006). Studies on the dual antioxidant and antibacterial properties of parsley (*Petroselinum crispum*) and cilantro (*Coriandrum sativum*) extracts. *Food Chemistry, 97*(3), 505-515.

Wood, J.D., Enser, M., Fisher, A.V., Nute, G.R., Sheard, P.R., Richardson, R.I., Hughes, S.I., & Wittington, F.M. (2008). Fat deposition, fatty acid composition and meat quality: A review. *Meat Science, 78*(4), 343–358.

Xiong, G., Wang, P., Zheng, H., Xu, X., Zhu, Y., Zhou, G. (2016). Effects of plant oil combinations substituting pork back-fat combined with pre-emulsification on physicochemical, textural, microstructural and sensory properties of spreadable chicken liver paté. *Journal of Food Quality, 39(4),* 331-341.

Zheng & Wang, (2001). Antioxidant activity and phenolic compounds in selected herbs. *Journal of Agricultural and Food Chemistry, 49*(11), 5165-5170.

In: Beef: Production and Management Practices ISBN: 978-1-53613-254-0
Editor: Nelson Roberto Furquim   © 2018 Nova Science Publishers, Inc.

Chapter 3

THE SUPPLY OF SAFE BEEF IN BRAZIL: A DISCUSSION ABOUT THE EFFICACY OF THE TRACEABILITY SYSTEM IMPLEMENTED IN THE PRODUCTION CHAIN

Nelson Roberto Furquim, PhD*
Universidade Presbiteriana Mackenzie, Brazil

ABSTRACT

Problems related to food safety, associated to animal health, such as the avian flu and the mad cow disease (Bovine Spongiform Encephalopathy – BSE), have gained prominence since the decade of 1990 and influenced not only the perception of consumers about the quality of animal based food products, but also the international trade policies in several countries. To meet the requirements of countries that started to demand traceability in the food production chains, in 2002, Brazil developed and implemented the Brazilian System for the

---

* Corresponding Author Email: nrfurquim@alumni.usp.br.

Identification and Certification of Bovine and Bubaline Origin – SISBOV - a system for monitoring the production chain as a whole, with mandatory accession for cattle ranchers exporting to countries with that requirement. Therefore, in this complex institutional environment, it is worth investigating the opinion of the economic agents involved in the beef production chain in Brazil about the effectiveness of SISBOV. A study was conducted based on the application of the qualitative and quantitative methodology of the Discourse of the Collective Subject (DCS), which allows identifying consensus about a theme or phenomenon based on the responses of a group of individuals. The following groups of economic actors involved in the cattle activity were intentionally defined for a nationwide survey, which constituted a non-probabilistic, convenience sample, by spontaneous adhesion: cattle ranchers, slaughterhouses members of the Brazilian Association of Meat Exporting Industries (ABIEC); companies certifying cattle ranches for export; associations and class entities directly related to actors in the beef production chain; Brazilian government agencies focused on food inspection and control. The study has evaluated the beef production chain actors´ opinion about SISBOV efficacy. For that purpose, the methodology of the Discourse of the Collective Subject (DCS) was applied to a sample of 34 actors of the production chain. From the interviews conducted for the study it was evidenced that there is resistance to accession to the traceability system mainly because of the significant investments required from cattle ranchers, with neither guarantee of overprice for the traced cattle nor financial return. Respondents indicated that the system is targeted only at markets requiring traceability; it is subject to the influences of international trade and political agreements, besides being ineffective due to regulatory failures. Those aspects demonstrate the dissatisfaction of the players in this market in relation to the expectations they had about SISBOV. Respondents also criticize government coordination, considered inefficient and with structural deficiencies in the control and regulatory services. From the perspective of possible benefits brought by SISBOV, comments are made about its contribution for more effective management of herds and ranches in the case of accession. Comments from the respondents suggest that they expected that SISBOV would promote coordinated actions among all elements of the meat agribusiness chain. Thus, the proposal of greater articulation among the members of the productive chain of bovine meat in the country arises, including the government at all levels, to address improvements and adjustments to the SISBOV design, promoting the coordination of all links within the production chain.

Keywords: supply, beef, traceability system, production chain, food safety

INTRODUCTION

Problems related to food safety, associated to animal health, such as the avian flu and the mad cow disease (Bovine Spongiform Encephalopathy – BSE), have gained prominence since the years 1990 and influenced not only the perception of consumers about the quality of animal based food products, but also the international trade policies in several countries. The European Union (EU), for instance, created a series of requirements to be met by products imported by the region, due to the international episodes that involved the commercialization of contaminated beef (Mendes, 2006). To meet those requirements, Brazil developed and implemented the Brazilian System for the Identification and Certification of Bovine and Bubaline Origin – SISBOV - a system for monitoring the production chain as a whole (BRASIL, 2002a; Mendes, 2006; Velho et al., 2009).

In fact, problems of food contamination derived from livestock raw materials are very difficult to treat and control, taking into account the emergence of new pathogenic microorganisms, in addition to those already known, with broad scope of action and resistance to antibiotics. That requires dynamic monitoring systems and the involvement of national and international agencies in food inspection (Sofos, 2008 and 2009).

There are several government agencies that monitor and control eventual food contamination problems, such as the Food and Drug Administration (FDA), World Health Organization (WHO), Codex Alimentarius Commission (CCA), United Nations Food and Agriculture Organization (FAO), National Health Surveillance Agency (Anvisa, in Brazil), among others. They are continuously searching for refined and rapid methods for detection of pathogenic microorganisms and for control of foodborne diseases (FBDs), by means of risk assessment and control of critical points in production processes.

Unlike bacterial contamination, Bovine Spongiform Encephalopathy (EBB), known as the "mad cow disease", originates from ranches, where healthy cattle may be fed with animal food prepared with meat and bones

72 Nelson Roberto Furquim

from contaminated livestock, so that it becomes a vector of contamination (Nardone, 2003).

In those cases, the existence of a traceability system (TRS) that allows the tracing back and forth in the production chain, including the origin of used inputs, reduces the likelihood of infected meat and meat based products reaching the end consumer.

In addition to that aspect of public health, a TRS, by increasing the flow of information along the chain, reduces the incidence of opportunistic behavior among actors in the sector, increasing competition, with impact on prices (Rubin, Ilha, Waquil, 2008), and in the international trade. In this sense, a TRS is a mechanism with which to deal with non-tariff barriers based on the requirement of traceability (Brown et al., 2001).

In Brazil, accession to the Brazilian System for the Identification and Certification of Bovine and Bubaline Origin (Sisbov), which is mandatory for cattle ranchers exporting to countries requiring traceability, on one hand implies the incorporation of a set of controls and information technology in all links of the beef productive chain, increasing the complexity of the cattle activity management and its costs within the scope of each agent. On the other hand, it demands nationwide coordination and control, taking into account the size of the herd and its geographical distribution, given the continental dimensions of the country (Lopes, Santos, 2007).

According to the latest agribusiness census (Brasil, 2006b), accession to SISBOV is still very low, suggesting that, even though theoretically the Brazilian system is adequate, in practical terms it faces difficulties in achieving the objectives it proposes, such as breaking the non-tariff barriers of the EU against Brazilian beef.

Therefore, it becomes important to evaluate the opinion of the economic agents involved in the national beef production chain and in the export of its products about the operability and functionality of Sisbov, seeking to understand the reasons for the low accession of farmers to the system, which is a gateway to the international market.

In fact, from the late 1990s, with the impact caused by diseases such as BSE, transmitted to humans through the consumption of contaminated beef

(Nardone, 2003; Machado, Nantes, 2004), governments in several countries started to adopt standards and impose stricter requirements for the international commercialization of this food (Cócaro, Jesus, 2007), impacting farmers, intermediate traders and public agencies, to ensure a safe supply for human consumption (Spers, 2003).

The EU, seeking to protect its consumers, started to require the traceability of local and imported cattle by the region (Machado, Nantes, 2004; Monteiro, Caswell, 2004; Resende Filho, 2008), creating a non-tariff barrier to the Brazilian exports and to the ones from other countries. Thus, aiming at guaranteeing the quality of the supplied foods by means of the characterization of the origin, the sanitary status records and the productive protocols, the TRS have become an essential condition for continuous and safe access to international markets (Brown et al., 2001).

As pointed out by Furquim and Cyrillo (2012), the EU is an important market for Brazilian beef exports, and, due to the requirements of that trading bloc, there was a need to implement a TRS in the Brazilian production chain, under the coordination of the government (Silva, Batalha, 2000), implying higher production costs (Pitelli, Moraes, 2006; Velho et al., 2009).

In 2002, by means of Normative Instruction (NI) n. 01, of January 9, the Ministry of Agriculture, Livestock and Food Supply (MAPA) instituted Sisbov (Brasil, 2002a; Mendes, 2006), in order to meet the EU traceability requirements for imports of Brazilian beef (Sarto, 2002).

Designed to ensure the accreditation of Brazilian beef exports to the EU, SISBOV includes in its structure a database with detailed information about the herd, how it is handled, and its movement in the national territory (Brasil, 2002a).

SISBOV was implemented with the support of the Agricultural Defense Secretariat (SDA) of MAPA, responsible for managing the National Data Base (BND) and for the procedures for accrediting the ranches certifying entities, producers and involved cattle (Brasil, 2002a).

With the implementation of SISBOV, the guidelines for traceability in the bovine and buffalo production chain were established, aiming at the registration and identification of herds, allowing the cattle to be tracked from birth to slaughter. The system adhesion is voluntary to the domestic market, being mandatory for beef and buffalo exports to markets that require traceability in the productive chain (Brasil, 2002a).

All animals registered in SISBOV must have an identity document, named Animal Identification Document (DIA), issued by a certifying entity, containing the ranch of origin, each animal identification, gender, date of birth or arrival at a given farm, feeding and breeding system, control and recording of movements (sales, purchases, deaths and transfers to other ranches), besides sanitary data, such as treatments, vaccines, among others (Cócaro, Jesus, 2007). Within SISBOV, the DIA permanently certifies the origin of the cattle, individually identified and registered in the National Data Base (BND), proving that they come from legally established ranches, rural properties, favoring their characterization and monitoring throughout the national territory.

On July 14, 2006, through the NI n. 17, MAPA presented a new operational structure for SISBOV, stressing that the adhesion to the Traceability Service of the Bovine and Bubaline Production Chain, the New SISBOV, is voluntary for cattle ranchers in general, but it is mandatory for those who are interested in exporting beef and buffalo to countries that require traceability in the production chain (Brasil, 2006c).

With that NI, which maintained the acronym SISBOV to designate Traceability Service of the Bovine and Bubaline Production Chain, emerges the concept of SISBOV Approved Farm (ERAS), whose main requirements and characteristics are: farm register, farmer register, basic production protocol, term of adhesion to SISBOV, recording of the inputs used in the property, individual identification of 100.0% of the bovines and buffaloes of the ranch, control of cattle movement, supervision of a single certifier accredited by MAPA and periodic inspections by the certifying entity (Brasil, 2006c; Cócaro, Jesus, 2007).

Cócaro and Jesus (2007) also mention that, with the New SISBOV, all bovines and buffaloes born in ERAS will be, obligatorily, identified individually before the first movement, in the period of time between weaning and, at the most, ten months of life. Registrations of those animals are made at the BND, being also necessary the registration of all the inputs

The Supply of Safe Beef in Brazil 75

used in the property during the productive process, which must be maintained for a period of five years.

According to NI n. 65 of December 16, 2009, SISBOV started to receive the designation of System for Bovine and Bubaline Identification and Certification, with the same acronym (Brasil, 2009a).

Even with the traceability system implemented in the beef productive chain for export, in 2005 there was an embargo of Brazilian exports to the EU and other countries, due to the outbreaks of foot-and-mouth disease in some Brazilian regions (Agronotícias, 2005). This event could be seen as an indication of ineffectiveness of the agricultural and cattle defense measures and the SISBOV, if it were not for the fact of the low adhesion of the cattle ranches to the referred system, as shown by the 2006 Census of Cattle and Agriculture (Brasil, 2006b; Furquim, Cyrillo, 2012).

In March 2017, dozens of federal inspectors were arrested after an investigation named "Weak Flesh", due to the acceptance of bribes to ignore the adulteration or expiration of processed meat-based products and falsified sanitary permits, which involved food-processing companies of different sizes all over Brazil (Le Monde, 2017; The New York Times, 2017). It is worth mentioning that since its implementation, the accession to SISBOV has been quite low in Brazil. In January 2016, there were 1,640 ranches authorized for export to the EU (Safra, 2016). In 2012, 1,948 ranches were accredited for export to that trading bloc (Correio Do Estado, 2012).

Thus, in this complex institutional environment - which seeks to manage the market failures inherent to the beef production and export sectors - it is worth investigating the opinion of the economic agents involved in the beef production chain in Brazil about the effectiveness of SISBOV, and about the low accession of the national cattle ranchers to the system.

The objective of the study was to investigate the opinion of cattle ranchers, slaughterhouses members of the Brazilian Association of Meat Exporting Industries (ABIEC), companies certifying cattle ranches for export, associations and class entities directly related to actors in the beef production chain, and Brazilian government agencies focused on food

76   Nelson Roberto Furquim

inspection and control, about government measures and services aimed at safe beef production, mainly regarding SISBOV.

METHODOLOGY

The current study used the qualitative and quantitative methodology of the Discourse of the Collective Subject (DCS), which allows the identification of consensus about a theme or phenomenon from the responses of a group of individuals. With this methodology, the answers obtained are classified through the recognition of Key Expressions (KE) and Central Ideas (CI), identified in the processing of the statements and used in the elaboration of speeches representative of the ideas of the group (Lefèvre and Lefèvre 2003).

The following groups of economic actors involved in the cattle activity were defined, intentionally, for a nationwide survey, and they constituted a non-probabilistic, convenience sample, by spontaneous adhesion: cattle ranchers, slaughterhouses members of the Brazilian Association of Meat Exporting Industries (ABIEC), companies certifying cattle ranches for export, associations and class entities directly related to actors in the beef production chain, and Brazilian government agencies focused on food inspection and control, such as the MAPA and the Ministry of Health (MS). Respondents who did not belong to the selected categories were excluded.

Electronic mail was used to send out invitations to take part in the research, together with the information and guidelines for answering the questionnaire using the QLQT Online software, version 1.0. The survey was conducted during October and November 2011. For this study, the answers to a specific question about the actors' knowledge about the effectiveness of SISBOV were analyzed.

The study was approved by the Ethics in Research Committee of the Pharmaceuticals Sciences College of the University of São Paulo according to Parecer CEP/FCF/146/2011, Protocolo CEP/FCF/595, CAAE: 0043.0.018.000-11.

RESULTS

The study was based on a sample of 34 individuals (about 8.0% of all invitations sent), belonging to the five different categories of selected actors and related to the beef cattle activity in Brazil. Among the respondents, 85.0% were male, 47.0% were cattle ranchers, and 97.0% had, at least, attended college, indicating a high level of education, being most of them veterinarians/zootechnicians (41.0%), followed by engineers (24.0%).

From the answers to the proposed question, five CI categories were established, as it can be observed in Table 1.

The proposed question aimed at identifying, in the perception of the respondents, whether SISBOV is fulfilling its role of enabling the export of Brazilian beef. Taking into account the purpose of the study, the question allows considering whether, in practice, SISBOV is consistent with its objectives and whether it is effectively operating in Brazil.

Among the answers to the question, 48.0% of the total indicates that SISBOV is not fulfilling its role. The system is considered expensive, ineffective, with no satisfactory results, mainly in terms of financial returns to cattle ranchers. The slaughterhouses are the ones that establish the values to be paid to the ranchers, without any guarantee of better prices for the traced cattle.

The respondents emphasize that SISBOV fundamentals are appropriate in its design, but its operation does not keep up with them. It is further believed that if it were fulfilling its role, there would be no additional imposition by EU government agencies.

There were also responses (10.0% of the total) that mentioned that SISBOV only partially fulfills its role, since there are failures in its operation. According to the respondents, the points to be adjusted - systematization and frequency of audits by the government - would contribute to achieving the objectives of the system, in the case of beef exports and cattle sanitary control.

Table 1. Distribution of the frequency of CI categories from the discourses of different actors in the Brazilian beef production chain, from the question "Do you think SISBOV is fulfilling its role? Please, make a few comments about it" – Brazil, 2011

Category		Answers	%
A	SISBOV is not fulfilling its role	20	47,6
B	SISBOV is partially fulfilling its role	4	9,5
C	SISBOV is fulfilling its role	9	21,4
D	Suggestions for improving SISBOV	8	19,0
E	No opinions about SISBOV being or not fulfilling its role	1	2,4
	Total	42	100,0

Source: the author, 2017.

On the other hand, about 21.0% of the answers to the question indicate that SISBOV is fulfilling its role. According to the respondents, the system meets the EU international requirements, and the cattle ranchers who choose to join it benefit from its practices to improve their routines in managing the herds of their ranches.

Among the suggestions for improving of SISBOV, observed in 19.0% of the responses, the respondents mention that it should be constantly improved, at the same time that there should be greater awareness of the producers and financial support from the government. Still according to the answers, the MAPA would have to supervise the performance of the certifying companies and ranches so that SISBOV could achieve greater international recognition.

DISCUSSION

In Brazil, the institutional framework focused on food safety encompasses both basic norms (Brasil, 1969) and the National System of Sanitary Surveillance, conducted by ANVISA (Brasil, 1999) to control and supervise the production, distribution and commercialization of food. The rules that regulate food supply and safety in Brazil are consistent with international standards (Furquim, Cyrillo, 2012), including labeling (Brasil, 2002b e 2005), which, even without requiring full traceability, make it mandatory the complete food identification information, including nutritional aspects and the names of manufacturers.

Both MAPA and ANVISA hold responsibility for the technical standards of sanitary safety of foods supplied in Brazil, even of beef, constantly reviewing them, and seeking their improvement and the guarantee of safe food. In particular, SISBOV was implemented as a response to the EU requirements, due to the problems generated by the "mad cow" epidemic (Moe, 1998; Golan et. al., 2004; Bennet, 2008).

The system defines the rules for registration and control of cattle for export, similar to the rules created by the EU for internal transactions. Those rules, which seek to monitor the cattle movement and inputs used in production, in fact are suitable for confined breeding. However, in Brazil, where such technology is not usual, and cattle are raised freely, in large areas of pasture, naturally fed; traceability, besides having difficult operation, is inefficient.

Although MAPA (Brasil, 2002a) has made the service available and outlined the necessary requirements to ensure traceability, the necessary investments are a responsibility of the cattle ranchers themselves, leading to additional costs in their activities (Cócaro, Jesus, 2007; Ventura, 2010), without improving the quality of the finished product. That is a critical aspect, which compromises the scope and effectiveness of the system in Brazil. On the one hand, there is the extent and dispersion of production in all geographic areas of the country, and, on the other hand, there are the investments to be made for the implementation of the TRS on the ranches.

From the interviews conducted for the study, it was evidenced that there is resistance to accession to this system mainly because the investments required from the cattle ranchers are significant, with neither guarantee of overprice for the traced cattle nor of financial return. Consequently, the number of traced cattle in the country is small, and the accession to SISBOV is basically observed among the most capitalized cattle ranchers.

Another critical aspect is pointed out by the research participants: besides the investments to be made on the ranches to implement SISBOV in the productive chain, producers also resist joining the system because of the specific nature of those investments, which become irrecoverable in the case of giving up production for export, or in downturns in beef exports, characterizing the "sunk costs" effect, since the domestic market does not pay overprice for traced cattle.

Despite the importance of beef for the national economy: 1.4 million tons or US$ 5.5 million in 2016 (Abiec, 2017), the high degree of complexity in its productive chain shows a situation of information asymmetry among the different participants, specially between cattle ranchers and slaughterhouses (Urso, 2007), an asymmetry that has been minimized in the perspective of the slaughterhouses and accentuated in the perspective of the ranchers, who, with the implementation of SISBOV, became hostages to the buying agent of their product.

In this sense, some respondents perceive SISBOV as a system that does not increase the transparency in the transactions among the different actors of the productive chain. On the contrary, it is understood as a strict inspection system, which limits and exposes the actions of producers, restricting any opportunistic behavior (such as management failures, clandestine slaughtering, among others), that could arise from them, due to information asymmetries.

In that context, the large number of ranchers, spread all over the country, and the reduced number of slaughterhouses (Brasil, 2006b) configure an oligopsone situation (Martins et. al., 2005; Golani, Moita, 2010), which was highlighted by the respondents, especially regarding the fact that the slaughterhouses are the ones responsible for establishing prices in the commercial transactions.

Another market failure in the beef production chain may be noticed in cases of not complying with the basic rules for cattle health guarantee. Although MAPA acts regulating the production both from the point of view of the cattle sanitary aspects and the control of the legal practices related to the quality of the supplied meat (Brasil, 2001; 2002a; 2007b), according to the respondents, there is still clandestine slaughtering and the supply of products with lesser quality.

Those episodes lead to a negative perception about SISBOV. Respondents indicated failures in its implementation, since it was not designed and implemented in a way consistent with the features of the national cattle industry, without taking into account the changes and adjustments in its regulatory aspects (Brasil, 2006c), which have contributed to generate discredit and distrust from the domestic producers and importers.

The embargo on the exports of Brazilian beef due to outbreaks of foot-and-mouth disease in 2005 (Agronotícias, 2005), and more recently, the "Weak Flesh" investigation (Le Monde, 2017a; The New York Times, 2017) appear as evidences of the system's ineffectiveness.

It should be noticed that in the national context uncertainties regarding the production and the delivery of the finished product may arise, besides information asymmetries between the intermediate agents (the slaughterhouses) and the inspection agents. That promotes the rising of the market and government failures, especially corruption.

That possibility was real with the police operation called "Weak Flesh Investigation", triggered in March 2017, against Brazilian slaughterhouses suspected of falsifying the quality of commercialized meat, both in the domestic market and for export, in complicity with MAPA agents (Le Monde, 2017a). In June 2017, new allegations of product adulteration led to an embargo on Brazilian exports of beef by the United States (Le Monde, 2017c).

Such illicit maneuvers severely endanger the credibility of Brazilian beef in the domestic and international markets, in detriment of the image of the sector and suspension of imports, leading to declines in export volumes and values (Le Monde, 2017b), which damages Brazil's position as the world's largest beef exporter (Le Monde, 2017c; The New York Times, 2017).

From this perspective, questions arise regarding ethics in the performance of sanitary inspectors, the effectiveness and validity of the

82 *Nelson Roberto Furquim*

supervision and control mechanisms implemented by Brazil to ensure the supply of safe food (El País, 2017).

Those aspects, however, do not solve the problems pointed out by the respondents in indicating that the system is focused only on markets that require traceability. It is also considered as subject to the influences of international trade and political agreements, besides being ineffective due to regulatory failures. Those features may indicate the dissatisfaction of the players in the sector, as far as the expectations they had about SISBOV are concerned.

Respondents also criticize government coordination, which is considered to be inefficient and presenting design failures in the cattle sanitary control and defense services.

From the perspective of possible benefits brought by SISBOV, there were comments about its contribution to a more effective management of herds and of the ranches, in cases of accession. However, the respondents' comments suggest that they had higher expectations about the system. They expected, for instance, the system to promote coordinated actions among all elements of the meat agribusiness chain, and that it should not only apply to ranches.

Thus, the proposal of greater articulation among the members of the productive chain of bovine meat in the country arises, including the government at all levels, to address improvements and adjustments to the SISBOV design, promoting the coordination of all links within the production chain.

CONCLUSION

The Brazilian agribusiness has traditionally been important for guaranteeing surpluses in the trade balance and for the inflow of foreign currency into the country. In that context, beef plays a significant role because besides having a proven nutritional importance, it has a high share in the value of total Brazilian exports.

The Supply of Safe Beef in Brazil

SISBOV, whose accession is voluntary for those not aiming at exporting Brazilian beef, is perceived as necessary only for international markets requiring traceability, while the domestic market is still considered undemanding.

Actors in the beef production chain perceive it as an ineffective system that, even being coordinated by the government, presents private features, being financed by the cattle ranchers themselves.

In 2009 the transparency law was implemented in Brazil (Brasil, 2009), which allows access to public information, but this law still lacks effectiveness (FGV, 2017). However, data on the slaughterhouses market and inspection information are practically nonexistent. There is a need for a policy that can favor the availability and quality of the information relevant to the parties involved.

In addition, it should be mandatory to prioritize the enforcement capacity and ethics of the Brazilian police and judicial authorities to inhibit the opportunistic behavior of private and public agents, increasing the efficiency of the beef market and the food markets in general.

It is necessary to have trained and certified public agents not only in their technical knowledge, but also in their suitability.

REFERENCES

Agronotícias. *Febre aftosa: 41 países já anunciaram suspensão da compra de carne brasileira.* Brasília, 2005. [*Foot-and-mouth disease: 41 countries have already announced a suspension of the purchase of Brazilian beef.* Brasília, 2005]. Available at: <http://www.agroportal. pt/x/agronoticias/2005/10/21h.htm>. Access: 19 Nov. 2011.

Associação Brasileira Das Indústrias Exportadoras De Carne (ABIEC). *Exportações brasileiras de carne bovina.* 2017. [*Brazilian beef exports.* 2017.]. Available at: <http://www.abiec.com.br/download/ exportacoes-jan-dez-2016.pdf >. Access: 17 Jun. 2017.

Bennet, G. S. *"Identity preservation & traceability": the state of the art - from a grain perspective (status of agricultural quality systems/*

84   Nelson Roberto Furquim

traceability/certification systems). Thesis (PhD in Philosophy), Iowa State University, Ames - Iowa, 2008.

Brasil. *Presidência da República. Casa Civil. Decreto-Lei nº 986, de 21 de outubro de 1969.* Brasília, 1969. [Presidency of the Republic. Civil House. Law Decree nº 986, of October 21, 1969. Brasília, 1969.]. Available at: <https://www.planalto.gov.br/ccivil_03/decreto-lei/ Del 0986.htm>. Access: 15 Jan. 2011.

Brasil. *Presidência da República. Casa Civil. Lei nº 9.782, de 26 de janeiro de 1999.* Brasília, 1999. [Presidency of the Republic. Civil House. Law nº 9,782, January 26, 1999. Brasília, 1999.]. Available at: <https://www.planalto.gov.br/ccivil_03/leis/l9782.htm>. Access: 09 Apr. 2011.

Brasil. Ministério da Agricultura e do Abastecimento. Secretaria de Desenvolvimento Rural. Instrução Normativa nº 10, de 27 de abril de 2001. Dispõe sobre a proibição de importação, produção, comercialização e uso de substâncias naturais ou artificiais com atividade anabolizante, ou mesmo outras dotadas dessa atividade, mas desprovidas de caráter hormonal, para fins de crescimento e ganho de peso em bovino de abate e revoga a Portaria nº 51, de 24 de maio de 1991. *Diário Oficial da União. Brasília, April 30, 2001.* [Ministry of Agriculture and Supply. Secretariat of Rural Development. Normative Instruction nº 10 of 27 April 2001. Provisions on the prohibition of importation, production, commercialization and use of natural or artificial substances with anabolic activity, or even others endowed with that activity, but devoid of hormonal character, for purposes of growth and weight gain in slaughter cattle and revokes Portaria nº 51, of May 24, 1991. Federal Official Journal. Brasília, April 30].

Brasil. Ministério da Agricultura, Pecuária e Abastecimento. Gabinete do Ministro. Instrução Normativa nº 1, de 10 de janeiro de 2002. Institui o Sistema Brasileiro de Identificação e Certificação de Origem Bovina e Bubalina - SISBOV. *Diário Oficial da União.* Brasília, seção 1, p.6, 10/01/2002a. [Ministry of Agriculture, Livestock and Supply. Minister's Office. Normative Instruction nº 1 of January 10, 2002. Institutes the Brazilian System for the Identification and Certification

The Supply of Safe Beef in Brazil 85

of Bovine and Bubaline Origin – SISBOV. Federal Official Journal. Brasília, section 1, p.6, 10/01/2002a].

Brasil. Ministério da Saúde. Agência Nacional de Vigilância Sanitária - ANVISA. *Resolução - RDC n° 259, de 20 de setembro de 2002.* Brasília, 2002b. [Ministry of Health. National Health Surveillance Agency – ANVISA. Resolution - RDC n° 259 of September 20, 2002. Brasília, 2002b]. Available at: <http://www.anvisa.gov.br/legis/resol/ 2002/259_02rdc.htm>. Access: 18 Apr. 2011.

Brasil. Ministério da Agricultura, Pecuária e Abastecimento. Gabinete do Ministro. *Instrução Normativa n° 22, de 24 de novembro de 2005.* Brasília, 2005. [Ministry of Agriculture, Livestock and Supply. Minister's Office. Normative Instruction n° 22, of November 24, 2005. Brasília, 2005]. Available at: <http://extranet.agricultura.gov. br/ sislegis-consulta/consultarLegislacao.do?operacao= visualizar&id= 14493>. Access: 18 Apr. 2011.

Brasil. Ministério do Planejamento, Orçamento e Gestão. Instituto Brasileiro de Geografia e Estatística - IBGE. *Censo Agropecuário 2006.* Rio de Janeiro, 2006b. [Ministry of Planning, Budget and Management. Brazilian Institute of Geography and Statistics - IBGE. *Agriculture and Livestock Census 2006.* Rio de Janeiro, 2006b]. Available at: <http://www.ibge.gov.br/home/estatistica/economia/ agropecuaria/censoagro/brasil_2006/defaulttab_brasil.shtm>. Acess: 29 May. 2011.

Brasil. *Ministério da Agricultura, Pecuária e Abastecimento. Cartilha do novo serviço de rastreabilidade na cadeia produtiva de bovinos e bubalinos – SISBOV* [Ministry of Agriculture, Livestock and Supply. Booklet of the new traceability service in the bovine and buffalo productive chain – SISBOV]. Brasília: SDC/ABIEC/CNA/ACERTA, 2006c.

Brasil. Ministério da Agricultura, Pecuária e Abastecimento. Gabinete do Ministro. *Instrução Normativa n° 44, de 02 de outubro de 2007.* Aprova as diretrizes gerais para a Erradicação e a Prevenção da Febre Aftosa, constante do Anexo I, e os Anexos II, III e IV, desta Instrução Normativa, a serem observados em todo o Território Nacional, com

86   *Nelson Roberto Furquim*

vistas à implementação do Programa Nacional de Erradicação e Prevenção da Febre Aftosa (PNEFA), conforme o estabelecido pelo Sistema Unificado de Atenção à Sanidade Agropecuária. Brasília, 2007b. [Ministry of Agriculture, Livestock and Supply. Minister's Office. Normative Instruction n° 44, of October 2, 2007. Approves the general guidelines for the Eradication and Prevention of Foot-and-Mouth Disease, contained in Annex I, and Annexes II, III and IV of this Normative Instruction, to be considered nationwide, aiming at the implementation of the National Program for the Eradication and Prevention of Foot-and-Mouth Disease (PNEFA), as established by the Unified System of Attention to Agricultural and Livestock Health. Brasília, 2007b]. Available at: <http://extranet.agricultura.gov.br/ sislegis-consulta/consultar Legislacao.do?operacao=visualizar&id= 18117>. Access: 08 Feb. 2012.

Brasil. *Confederação Nacional de Municípios. Lei da Transparência, 2009.* [National Confederation of Municipalities. Transparency Law, 2009]. Available at: http://www.leidatransparencia.cnm.org.br/. Access: 09 Aug. 2017.

Brasil. Ministério da Agricultura, Pecuária e Abastecimento. Gabinete do Ministro. *Instrução Normativa n° 65 de 16 de dezembro de 2009.* Altera a denominação do Serviço de Rastreabilidade da Cadeia Produtiva de Bovinos e Bubalinos - SISBOV, que passa a chamar-se Sistema de Identificação e Certificação de Bovinos e Bubalinos - SISBOV. Diário Oficial da União. Brasília, seção 1, p.19, 17/12/ 2009a. [Ministry of Agriculture, Livestock and Supply. Minister's Office. Normative Instruction n. 65 of December 16, 2009. Changes the name of the Traceability Service of the Production Chain of Bovines and Buffaloes - SISBOV, which is now called the System of Identification and Certification of Bovines and Buffaloes - SISBOV. Federal Official Journal. Brasilia, section 1, p.19, 17/12/2009a].

Brown, P.; Will, R. G.; Bradley, R.; Asher, D. M.; Detwiler, L. Bovine Spongiform Encephalopathy and Variant Creutzfeldt-Jakob Disease: Background, Evolution, and Current Concerns. *Emerging Infectious Diseases*, v.7, n° 1, p.6-16, 2001.

The Supply of Safe Beef in Brazil   87

Cócaro, H.; Jesus, J. C. S. Impactos da implantação da rastreabilidade bovina em empresas rurais informatizadas: estudos de caso. *Revista de Gestão da Tecnologia e Sistemas de Informação*, v.4, n. 3, p.353-74, 2007. [Impacts of the implementation of bovine traceability in computerized rural enterprises: case studies. *Journal of Technology Management and Information Systems*, v.4, n. 3, p.353-74, 2007].

Correio Do Estado. *MAPA divulga fazendas de MS aptas a exportar*. Seção Economia, 05/02/2012. [*MAPA discloses MS ranches suitable to export*. Economy Section, 02/05/2012. Available at: <http://www.correiodoestado.com.br/noticias/mapa-divulga-fazendas-de-ms-aptas-a-exportar_140525/>. Access: 08 Feb. 2012.

El País. *Operação carne fraca: o esquema podre que ronda os frigoríficos no Brasil*. 2017. [*Low meat operation: the rotten scheme that rounds the slaughterhouses in Brazil*, 2017]. Available at: https://brasil.elpais.com/brasil/2017/03/24/politica/1490391912_181027.html. Access: 02 Aug. 2017.

Fundação Getúlio Vargas (FGV). *FGV avalia os cinco anos da Lei de Acesso à Informação*. 2017. [*FGV evaluates the five years of the Law on Access to Information*, 2017]. Available at: http://portal.fgv.br/noticias/fgv-avalia-cinco-anos-lei-acesso-informação. Access: 4 Aug. 2017.

Furquim, N. R.; Cyrillo, D. C. Artigo 1 - Sistemas de Identificação e Rastreabilidade na cadeia produtiva alimentar: uma análise sob a perspectiva da oferta segura de carne bovina. In: Furquim, N. R.; Cyrillo, D. C. *"Alimento Seguro: uma análise do ambiente institucional para oferta de carne bovina no Brasil"*. [Article 1 - Identification and traceability systems in the food production chain: an analysis from the perspective of the safe supply of beef. In: Furquim, N. R.; Cyrillo, D. C. "Safe Food: an analysis of the institutional environment for beef supply in Brazil". Thesis (PhD in Applied Human Nutrition), Universidade de São Paulo, São Paulo, 2012.

Golan, E.; Krissof, B.; Kuchler, F.; Calvin, L.; Nelson, K.; Price, G. Traceability in the U.S. Food Supply: Economic Theory and Industry

Studies. U.S. Department of Agriculture, Economic Research Service *AER* 830, March 2004.

Golani, L.; Moita, R. O oligopsônio dos frigoríficos: uma análise empírica de poder de mercado [The oligopsone of slaughterhouses: an empirical analysis of market power]. Insper Working Paper. *WPE*: 228/2010. 2010. Available at <http://www.insper.edu.br/sites/default/files/2010 _wpe228_0.pdf>. Access: 10 Jan. 2012.

Le Monde. *Viande avariée: le Brésil face à des représailles.* 2017a. [*Damaged meat: Brazil faces reprisals*, 2017a. Available at: <http://www.lemonde.fr/economie/article/2017/03/21/viandeavariee-le-bresil-face-a-des-represailles_5098221_3234.html>. Access: 30 Jun 2017.

Le Monde. *Au Brésil, les autorités au secours de l'industrie de la viande.* 2017b. [*In Brazil, the authorities are helping the meat industry*]. Available at< http://www.la-croix. com/Economie/Monde/Au-Bresil-autorites-volent-secours-lindustrie-viande-2017-03-26-1200834829>. Acess: 02 Aug 2017.

Le Monde. *Washington suspend les importations de boeuf brésilien pour raisons sanitaires.* 2017c. [*Washington suspended imports of Brazilian beef for sanitary reasons*]. Available at: < http://www.lemonde.fr/ ameriques/article/2017/06/23/washington-suspend-les importationsde-b-uf-bresilien-pour-raisons-sanitaires_5149733_3222.html?xtmc= viande_bresilienne&xtcr>. Access: 01 Aug 2017.

Lefèvre, F.; Lefèvre, A. M. C. *Discurso do sujeito coletivo: um novo enfoque em pesquisa qualitativa (desdobramentos)* [*Discourse of the collective subject: a new approach in qualitative research (unfoldings)*]. Caxias do Sul: EDUCS, 2003.

Lopes, M. A.; Santos, G. Principais dificuldades encontradas pelas certificadoras para rastrear bovinos. *Revista Ciência e Agrotecnologia* [Main difficulties encountered by certifiers to track cattle. *Science and Agrotechnology Review*]. Lavras, v.31, n.5, p.1552-7, Sep./Oct. 2007.

Machado, J. G. C.; Nantes, J. F. D. A rastreabilidade na cadeia da carne bovina. *I Congresso luso-brasileiro de tecnologias de informação e comunicação na agro-pecuária.* Santarém - Portugal, 2004.

The Supply of Safe Beef in Brazil

[Traceability in the beef chain. *I Portuguese-Brazilian Congress of information and communication technologies in agro-livestock.* Santarém - Portugal, 2004]. Available at: <http://www. agricultura digital.org/agritic_2004/congresso/Seg_e_Qual_Alim_Rastreab/A_Ras treabilidade_na_Cadeia_Carne_Bovina.pdf>. Access: 19 Mar 2011.

Martins, R. S.; Rebechi, D.; Prati, C. A.; Conte, H. Decisões estratégicas na logística do agronegócio: compensação de custos transporte-armazenagem para a soja no estado do Paraná. *Revista de Administração Contemporânea.* [Strategic decisions in agribusiness logistics: transportation-storage cost compensation for soybeans in the state of Paraná. *Contemporary Administration Review*], v. 9, n. º 1, 2005. Available at: <http://www.scielo.br/scielo.php? pid=S1415-65552005000100004&script= sci_arttext>. Access: 08 Feb. 2012.

Mendes, R. E. O impacto financeiro da rastreabilidade em sistemas de produção de bovinos no Estado de Santa Catarina [The financial impact of traceability in cattle production systems in the State of Santa Catarina], Brasil. *Ciência Rural,* v.36, n. 5, p.1524-8, Sep.-Oct. 2006.

Moe, T. Perspectives on traceability in food manufacture. *Trends in Food Science and Technology*, v. 9, p.211-4, 1998.

Monteiro, D. M. S.; Caswell, J. A. "The economics of implementing traceability in beef supply chains: trends in major producing and trading countries." Working Paper. n° 2004-6. Amherst, MA: University of Massachusetts Amherst, Department of Resource Economics, 2004.

Nardone, A. Impact of BSE on livestock production system. *Veterinary Research Communications,* v. 27, Supplement 1, p.39-52, 2003.

Pitelli, M. M.; Moraes, M. A. F. D. Análise do impacto das variações institucionais européias sobre a governança do sistema agroindustrial brasileiro da carne bovina. *Rev. Econ. Sociol. Rural* [Analysis of the impact of European institutional variations on the governance of the Brazilian beef agroindustrial system. *Rural Sociological Economics Review*]. Brasília, v. 44, n. 1, p.27-46, Mar. 2006. Available at: <http://www.scielo. br/scielo.php?script=sci_arttext&pid=S0103-200 32006000100002 &lng=pt&nrm=iso>. Access: 28 May 2010.

Resende Filho, M. A. Potenciais benefícios do sistema de rastreabilidade animal dos EUA para o setor de carnes americano. *Rev. Econ. Sociol. Rural* [Potential benefits of the US animal traceability system for the US meat sector. *Rural Sociological Economics Review*]. Brasília, v. 46, n. 4, Dec. 2008. Available at: <http://www.scielo.br/scielo.php? script=sci_arttext&pid=S0103-2003200800 0400009&lng=pt&nrm= iso>. Access: 28 May 2010.

Rubin, L. S.; Ilha, A. S.; Waquil, P. D. O comércio potencial brasileiro de carne bovina no contexto de integração regional [The potential Brazilian beef trade in the context of regional integration]. *RESR*, v. 46, n. ° 4, p.1067-94, 2008.

Safra. *Pecuária – Boi rastreado,* 2016. [*Livestock – Traced cattle,* 2016]. Available at: http://admin.agenciar8.com.br/uploads/arquivos/ 2016/04 /27/62eb4e353e3f720c85c66d1c1002b5a3img.pdf. Access: 16 Jun 2017.

Sarto, F. M. *"Análise dos impactos econômicos e sociais na implementação da rastreabilidade na pecuária bovina nacional"* ["Analysis of economic and social impacts in the implementation of traceability in national cattle farming"]. Under graduation term paper (Agronomic Engineering), Universidade de São Paulo, Escola Superior de Agricultura "Luiz de Queiroz", Piracicaba, 2002.

Silva, C. A. B.; Batalha, M. O. (Coord.) *Estudo sobre a eficiência econômica e competitividade da cadeia agroindustrial da pecuária de corte no Brasil.* Brasília: IEL, CNA and SEBRAE, 2000. [*Study on the economic efficiency and competitiveness of the cattle breeding agroindustrial chain in Brazil.* Brasília: IEL, CNA and SEBRAE, 2000].

Sofos, J. N. Challenges to meat safety in the 21st century. *Meat Science*, v. 78, p 3-13, 2008.

Sofos. ASAS Centennial Paper: Developments and future outlook for postslaughter food safety. *Journal of Animal Science*, v. 87, p.2448-57, 2009.

Spers, E. E. *"Mecanismos de regulação da qualidade e segurança de alimentos"* ["Mechanisms for regulating food quality and safety"].

Thesis (PhD in Business Administration), Faculdade de Economia, Administração e Contabilidade, Universidade de São Paulo, 2003.

The New York Times. Brazil Meat Scandal Is Called 'A Punch in the Stomach', 2017. Available at: https://www.nytimes.com/2017/03/22/world/americas/brazil-meat-industry-scandal-exports.html?_r=0;. Access: 30 Jun 2017.

Urso, F. S. P. *"A cadeia de carne bovina no Brasil": uma análise de poder de mercado e teoria da informação* ["The beef chain in Brazil": an analysis of the market power and the information theory]. Thesis (PhD in Economics), Fundação Getúlio Vargas, São Paulo, 2007.

Velho, J. P.; Barcellos, J. O. J.; Lengler, L.; Elias, S. A.; Oliveira, T. E. Disposição dos consumidores porto-alegrenses à compra de carne bovina com certificação. *R. Bras. Zootec* [The willing of Porto Alegre consumers to purchasing certified beef. *Brazilian Zootechny Journal]*, Viçosa, v. 38, n. 2, Feb. 2009. Available at: <http://www.scielo.br/scielo.php?script=sci_arttext& pid=S1516-35982009000200025&lng =pt&nrm=iso>. Access: 28 May 2010.

Ventura, C. A. A. Da negociação à formação dos contratos internacionais do comércio: especificidades do contrato de compra e venda internacional. *Revista Eletrônica de Direito Internacional* [From negotiation to the setting of international trade contracts. *International Law Digital Review]*. v. 6, p.90-121, 2010. Available at: <http://www. cedin.com.br/revistaeletronica/volume6/>. Access: 7 May 2011.

In: Beef: Production and Management Practices ISBN: 978-1-53613-254-0
Editor: Nelson Roberto Furquim   © 2018 Nova Science Publishers, Inc.

Chapter 4

EXPLORING THE BRAZILIAN CONSUMER'S PERCEPTION ABOUT SODIUM CHLORIDE REDUCTION IN FRANKFURTER TYPE SAUSAGES: A QUALITATIVE AND QUANTITATIVE APPROACH

Maria T. E. L. Galvão[1,], Rosires Deliza[2], PhD and Marise A. R. Pollonio[1], PhD*

[1]Department of Food Technology, Faculty of Food Engineering,
University of Campinas, UNICAMP, Campinas, SP, Brazil
[2]Embrapa Agroindústria de Alimentos, Rio de Janeiro, RJ, Brazil

ABSTRACT

Excessive consumption of sodium chloride has been widely criticized by the medical and scientific communities; this ingredient is

---

[*] Corresponding Author Email: mtelgalvao@gmail.com.

94 Maria T. E. L. Galvão, Rosires Deliza and Marise A. R. Pollonio

directly related to increased blood pressure, one of the main factors associated with cardiovascular events. Studies have shown that most sodium in an industrialized diet is presented in the food at the moment of purchase, added by the industry or restaurants. According to the Brazilian Association of Food Industries, the contribution of processed foods to daily sodium intake is 25-30%, depending on the geographic region. Regardless of its origin, the average daily sodium chloride intake worldwide is around 9 to 12 g, which is above the level of 5 g/day recommended by the World Health Organization, and this recommendation is diminishing even more. Numerous initiatives have been done in several countries to reduce sodium intake, such as establishing targets to reduce the sodium of industrialized products in partnership with industries, education campaigns, and encouraging the consumer to read the nutritional information on labels, as well as public awareness regarding the harmful effects of excessive salt intake. In Brazil, agreements between the Ministry of Health and the private sector have established gradual sodium reduction targets by the year 2017 for various food categories, including emulsified meat products, such as bologna and sausage. As reported by Nielsen (2015), these products account for about 60% of the category of meat products, eaten either as a main dish or meat substitute in breakfast, afternoon snacks, and school meals. In this context, Brazilian consumers' perception of salt reduction, associated with processed meat products was investigated using qualitative (focus group) and quantitative (Kano modelling) approaches. Concerns about the intake of processed meat products have focused on the frequency of consumption rather than their experience. A pre-existing sensory experience was considered a critical factor for purchasing frankfurter rather than the nutritional information, and information about sodium levels provided on labels. Flavor and texture were the most relevant attributes. The best before date and tenderness were considered attractive attributes. The analysis of better and worse scores indicated that the best before date, smooth texture, no grittiness during chewing, tenderness and being well-seasoned were characteristics that could improve consumers' satisfaction. Despite knowing the harmful effects caused by high salt intake, consumers are unaware of the recommended intake levels. In this context, public awareness initiatives on an issue so important to health should be conducted. It is expected that this study could contribute to the beef industry to reformulate processed meat products and add value to consistent sodium reduction claims.

Keywords: low sodium meat products, focus group, quality items, sensory descriptors

1. INTRODUCTION

Sodium chloride is the main dietary source of sodium, and excessive sodium intake is directly related to the occurrence of hypertension, one of the risk factors for the development of cardiovascular and cerebral accidents (Frisoli, Schmieder, Grodzicki, & Messerli, 2012; He & MacGregor, 2003; He & MacGregor, 2009; Neal, 2014; Svetkey, et al., 1999), cancer, bone mass loss, kidney disease, metabolic syndrome and obesity (Cappucio, Kalaitzidis, Duneclift, & Eastwood, 2000; D'Elia, Rossi, Ippolito, Cappuccio, & Strazzullo, 2012). Studies have shown that most sodium in an industrialized diet is presented in food at the moment of purchase, added by the industry or restaurants (M Dötsch, et al., 2009). According to the Brazilian Association of Food Industries (ABIA, 2013), the contribution of processed products to daily sodium intake is 25-30%, depending on the geographic region. Regardless of its origin, the average daily sodium content worldwide is around 9-12g/dia (Despain, 2014; Tobin, O'Sullivan, Hamill, & Kerry, 2012) and in Brazil this reality is no different; according to the Brazilian Association of Food Industries, the average consumption is also close to these values (ABIA, 2013), which is above the level of 5 g/day recommended by the World Health Organization (WHO, 2015; Zandstra, Lion, & Newson, 2016).

Meat products are among the target categories for sodium reduction, as they contribute to about 20% of total salt intake (Weiss, Gibbs, Schuh, & Salminen, 2010). However, the reduction levels depend on a number of factors, including the nature of the product, chemical composition, and type of manufacturing process (Desmond, 2006; Knight & Parsons, 1988; Ruusunen & Puolanne, 2005). In Brazil, emulsified products such as sausages and bologna represent more than 50% of processed meat products, which makes them targets for sodium reduction, once they are consumed by the entire country's population, regardless of age and socioeconomic status (ACNielsen, 2015 personal communication).

Sodium reduction in meat products is a complex challenge, as this ingredient has other functions besides taste (Toldrà & Reig, 2011). It is responsible for extracting the myofibrillar proteins and increasing the ionic

96   *Maria T. E. L. Galvão, Rosires Deliza and Marise A. R. Pollonio*

strength in the meat matrix, improving the functional properties such as emulsification, gelation, and water/fat holding capacity (Desmond, 2006). In addition to conferring the characteristic salty taste, salt enhances the overall taste of foods and suppresses the perception of bitter taste (Mariska Dötsch, et al., 2009), besides promoting satiety and contributing to salivation (Kuo & Lee, 2014). Thus, other technological strategies have been investigated rather than the simple reduction of sodium chloride, which is aimed to minimize losses associated with sodium reduction, including the partial replacement of sodium chloride by other salts (Horita, Messias, Morgano, Hayakawa, & Pollonio, 2014; Horita, Morgano, Celeghini, & Pollonio, 2011), as well as use of flavor enhancers (McGough, Sato, Rankin, & Sindelar, 2012). These studies focused mainly on sensory and physical characteristics and have not investigated in detail the consumers` perception about sodium reduction issues in this product category, and the possible impacts of such reduction on the perceived quality of the product.

Two types of information are required to understand consumers' perception: the consumers' expectation before purchase and consumers' perception after experimentation and use. According to van Kleef, van Trijp, and Luning (2005), evaluation before tasting can be driven by different methodological approaches, and the choice of one or the other will depend on the purpose of the study, type of orientation (product or need), and the way of obtaining this information (directly or indirectly).

Decision processes and the factors that influence the purchase decision of the product may be investigated, thereby providing effective subsidies for the technical development of the product (Moskowitz, Beckley, & Resurreccion, 2006). Consumers look for products not just for the sensory attributes, but also to meet their needs, including health benefits, flavor, and convenience issues (Gutman, 1982; Myers & Shocker, 1981). Recently, studies have shown that the production and processing methods have also been shown to be relevant in the purchasing decision process (Barcellos, et al., 2010; Guerrero, et al., 2009; Shim, et al., 2011; Verbeke, Pérez-Cueto, Barcellos, Krystallis, & Grunert, 2010).

Exploring the Brazilian Consumer's Perception about Sodium ... 97

Qualitative techniques such as focus groups (Krueger & Casey, 2009) have been widely used and recommended when working on incremental product development aiming to understand consumer expectations and needs for both services and products. In food, depending on the approach used, it is possible to understand the consumers' demands with respect to the intrinsic characteristics of a product or product category. However, since this methodology cannot identify which requirements may produce greater satisfaction, other techniques for assessing the effect of each attribute on customer satisfaction have been proposed (Sauerwein, Bailom, Matzler, & Hinterhuber, 1996), including the technique developed by Kano, Seraku, and Tsuji (1984). In the Kano modeling, the attributes can be classified into six categories according to satisfaction/dissatisfaction level as follows: The attractive attribute (it surprises but its absence does not cause dissatisfaction), proportional attribute (satisfaction increases if the attribute is implemented), must-be attribute (its presence is expected, and its absence causes dissatisfaction), indifferent attribute (the absence or presence does not cause neither consumer satisfaction nor dissatisfaction), reversal attribute (its presence can cause consumer dissatisfaction) and questionable. Besides this classification, Kano methodology allows for identifying the satisfaction and dissatisfaction levels, using "better" and "worse" scores, respectively, for each attribute, which are indicative of the effect of the attribute on consumers' satisfaction.

In this context, the present study aimed to investigate the consumer perception and involvement towards sodium reduction in sausages, and to identify the characteristics that impact more strongly on customer satisfaction among all investigated descriptors.

2. MATERIAL AND METHODS

2.1. Focus Group Session

Three focus group sessions were performed with female consumers in the city of Campinas SP, Brazil, belonging to socioeconomic classes A/B

98   Maria T. E. L. Galvão, Rosires Deliza and Marise A. R. Pollonio

(medium/high income classes), in three different age groups (25-35 years, 36-50 years, and above 50 years old). A fourth group consisting entirely of men from different ages and a similar socioeconomic level was also performed. The study was approved by the Ethics in Research Committee of the University of Campinas under number 385.996/2013 and was conducted in June of 2014.

Eight individuals participated in each discussion, totaling 32 people. Consumers were recruited from a database of an institute responsible for the recruitment. It was carried out by telephone and consumers answered a short questionnaire about food consumption habits, including their thoughts on the reduction of sugar, fat, calories, and salt, as well as the inclusion of dietary fiber and balanced diet. In addition, to participate in the study, the individuals should have eaten processed meat products such as sausage and bologna in the last fortnight. None of the participants worked directly with food or had prior knowledge about the subject to be discussed in the sessions.

Table 1. Interview guide used to perform the focus group sessions

Step	Description
1	Presentation of moderator and group dynamics. Welcome reception to participants
2	Introduction to the themes: food and healthiness
	• Food consumption at home, on weekdays and weekends;
	• Healthiness: definition; how to get information; how consumers deal with the subject at home
3	Consumption of meat and processed meat products
	• Types of meat consumed - consumption occasions, buying criteria, and preparation
	• Relationship between processed meat products and healthiness
	• Sausage 's purchase and consumption criteria (expected sensory characteristics, quality criteria when buying sausages)
4	Salt in processed meat products
	• Concerns about the subject, experiences and suggestions for sodium reduction
5	Degustation of products and finalization

Exploring the Brazilian Consumer's Perception about Sodium ... 99

An interview guide was developed and applied in all focus group sessions. It was focused on four subjects: Feeding in everyday life, perception of healthy food, consumption of meat and meat products, and level of perceived saltiness of meat products, as proposed by Krueger and Casey (2009) and shown in Table 1.

After Step 4 presented in Table 1, sausages with different sensory characteristics were presented to consumers, and they were asked to identify the sensory attributes that characterized the products. The detailed information of the sausages' formulation is shown in Section 2.2. In each discussion session, the products were presented in a randomized order, and consumers were advised to drink water and eat cracker biscuits for palate cleansing between samples. During tasting, consumers registered their perceptions of each product in an evaluation form, consisting of a blank page containing the product code (three-digit code). They were told to describe the perceived sensory characteristics rather than those related to acceptance/liking, thereby avoiding hedonic terms. After each tasting session, consumers shared their perceptions with the group.

Each discussion session lasted about an hour and a half. The same moderator led all interviews, accompanied by an assistant. All discussions were recorded and transcribed for analysis of results, as suggested by Bryman (2012).

2.2. Processing of Sausages for Eliciting the Sensory Attributes

To identify the sensory attributes, four sausages with distinct sensory characteristics were served to consumers aimed at facilitating attribute elicitation, with the results as follows: One sausage contained 2.5% sodium chloride, two low-salt sausage formulations (50% reduction) containing flavor enhancers, and a commercial formulation – a sales leader in the Brazilian market – containing 2.1% sodium chloride, were determined according to Latimer and Horwitz (2005). The flavor enhancers, lysine + ribonucleotides inosinate and disodium guanylate (IMP:GMP) and yeast extract + autolyzed yeast were defined in preliminary studies. The

100 Maria T. E. L. Galvão, Rosires Deliza and Marise A. R. Pollonio

formulations are presented in Table 2. The sausages were produced by a conventional process in a cutter and stuffed into 22 mm casings, then gradually cooked in an oven until the internal temperature reached 72°C. After cooking, the casings were peeled off and the sausages were placed in annatto solution. The products were vacuum-packed and stored at 4 ± 2°C for later evaluation within a maximum period of 15 days.

2.3. Quantitative Research

A quantitative research was carried out to evaluate the consumer's perception of each quality attribute previously identified in the focus group sessions, according to the methodology proposed by Kano et al. (1984). According to this methodology, a specific questionnaire is used to classify each quality attribute as indifferent, expected, proportional, attractive, reverse or questionable based on its presence (functional form of the question) or absence (dysfunctional form of the question) (Sauerwein, et al., 1996). One hundred and twenty consumers of sausages (65% female) from the socioeconomic classes A/B/C (high, medium and low/medium income respectively) according to the ABEP criteria (2013) were recruited for the study, which was carried out in Campinas, SP in October of 2015. For each quality attribute, a pair of questions was formulated. The first question was related to the consumers' reaction if the product had the attribute (the functional form of the question). The second one was related to his/her reaction if the product did not have the characteristic (non-functional form of the question). Consumers' reactions were evaluated in structured five-point scales varying from 1: I am very satisfied; 2: I am satisfied; 3: I am indifferent; 4: I am unsatisfied; and 5: I am very unsatisfied. Each pair of responses for the functional and non-functional questions was classified into six attributes, according to a reference table presented in Table 3 (Sauerwein, et al., 1996).

Exploring the Brazilian Consumer's Perception about Sodium ... 101

Table 2. Formulations used in the focus group study (%w/w)

Ingredients	Formulations (% w/w)		
	F1	F2	F3
Beef	40.00	40.00	40.00
Pork	20.00	20.00	20.00
Pork back fat	20.00	20.00	20.00
Ice	14.73	13.95	13.95
Salt	2.50	1.25	1.25
Phosphate	0.25	0.25	0.25
Sodium nitrite	0.02	0.02	0.02
Condiments	0.50	0.50	0.50
Maize starch	2.00	2.00	2.00
Lysine	--	2.00	--
IMP: GMP	--	0.03	--
Yeast extract	--	--	1.00
Autolyzed yeast	--	--	0.30

Table 3. Kano evaluation table

			Non-functional (negative question)				
			Very satisfied	Satisfied	Indifferent	Unsatisfied	Very unsatisfied
Functional (positive question)	questionattribute)	Very satisfied	Q	A	A	A	P
		Satisfied	R	I	I	I	E
		Indifferent	R	I	I	I	E
		Unsatisfied	R	I	I	I	E
		Very unsatisfied	R	R	R	R	Q

Attractive; P – Proportional; E – Expected; I – Indifferent; R – Reverse; Q – Questionable.

Frequency data were collected and the final classification of each attribute was defined according to the following criteria:

If $(\Sigma A + \Sigma E + \Sigma P) > (\Sigma R + \Sigma Q + \Sigma I)$, the classification is given by the greater frequency among A, E, and P;

If $(\Sigma A + \Sigma E + \Sigma P) < (\Sigma R + \Sigma Q + \Sigma I)$ the classification is given by the greater frequency among P, E, and I;

102 *Maria T. E. L. Galvão, Rosires Deliza and Marise A. R. Pollonio*

If $(\Sigma A + \Sigma E + \Sigma P) = (\Sigma R + \Sigma Q + \Sigma I)$ the classification is given by the one that respects the following order: $E > P > A > I$.

Better and *worse* scores were also calculated according to the following equations:

$$Better = (A + P)/(A + P + E + I)$$
$$Worse = (E + P)/(A + P + E + I)$$

These scores indicate how strongly a product characteristic may influence satisfaction or, in case of its "non-fulfillment", how strongly such a characteristic may influence dissatisfaction. The closer the value is to 1, the higher the influence on customer satisfaction. On the other hand, when it approaches 0 it means very little influence (Sauerwein, et al., 1996).

In addition to the Kano questionnaire, information about the consumers' perception of the harmful effects of high sodium intake to health, as well as the knowledge on the maximum recommended daily sodium chloride intake, attitudes towards salt intake and healthy eating habits were also collected in a separated questionnaire. Consumers answered their perception towards seven health disorders that could be associated to high salt intake using the most appropriate response among the following options (for sure, maybe, certainly not, and I do not know). In relation to the maximum recommended daily sodium chloride intake, consumers had to choose one of the five options (3g, 5g, 6g, 12g and don't know). Finally, issues related to attitudes towards healthy eating habits (e.g., use of natural products, without preservatives and processed food) and salt intake (salt eating habits and low salt diet) were answered using a 7-point scale (1: strongly disagree, 2: disagree, 3: disagree slightly, 4: neither agree/nor disagree, 5: somewhat agree; 6: agree; 7: strongly agree). Data were grouped into three categories: disagreement, indifference, and agreement, as described by Grimes, Riddell, and Nowson (2009) and analyzed using descriptive statistics and a Chi-square analysis for k-proportions to determine significant relationships between categories. All statistical analyses were conducted using the XLSTAT program (2014).

3. RESULTS

Consumers' responses obtained in the recruitment mentioned sliced ham, bologna and sausages as the most consumed meat products, the latter having higher frequency of consumption than the others (weekly). The majority of women reported consuming ham and bologna more than once a week, while among men, the highest consumption occurred weekly. The great consumption of ham and bologna among women is probably due to the fact that these products are widely used in the preparation of breakfast and afternoon snacks, or as a replacement for meals (especially in families with teenagers), or in school snacks.

All consumers also reported they have changed their habits in recent years. Women were more concerned about fiber inclusion in daily diet, physical activities, and balanced daily diet than men, who showed greater concern with sugar reduction. Items such as fat, calories, and salt reduction showed very similar results between genders, and fat reduction was the main concern of both groups.

Salt reduction was also relevant, being mentioned by most respondents, which proved to be a subject present in the minds of consumers.

Despite the reduced number of participants who participated in this step due to the qualitative characteristic of the study, the findings were useful to figure out the profile of consumers.

3.1. Focus Group

During the discussion sessions, concerns about healthiness were expressed in different ways according to the age group. In the group of younger consumers (25-35 years), healthiness was related to a natural diet rich in fruits and vegetables, preferably organic and with little use of industrial products. When having children younger than five years of age, the consumers of this group assumed the role of childcare providers and educators, thus the consumption of manufactured goods was less frequent, but with no restrictions. For mothers, the important thing is to eat fruits and vegetables every day, excluding unhealthy dietary products, as can be seen in some quotes below.

"At home we have a very healthy diet, but we eat everything we like. Once or twice a week we eat a sandwich at dinner - we like sausage, bologna, etc., but we try to eat healthy food as well". Woman, 33, one child

"We do not eat fried foods at home, and my son likes vegetables. He also enjoys sausage, but we try to restrict consumption on weekends. Regarding salt intake, continuous aging and its implications with unhealthy eating habits worries us". Woman, 35, one child

In the group of consumers aged 36-50 years old, healthiness was also mentioned through a balanced diet of fruits and vegetables. However, processed foods were more often present on the family menu. Familial meals at the table became rare, contributing to the consumption of processed meat products, mainly because children grew up and daily activities increased, as shown in the statement below:

"At home, we eat sliced meat products and bread - my children study at night so I have to offer these foods when they return from school". Woman, 45, two teenage children.

Consumers over the age of 50 years old believe that healthiness is related to a balanced diet and try to eat fresh foods. They also consider the influence of their adult children, who regularly understand nutritional responsibility and have a greater concern for health.

"My son is very concerned about health; he eats carbohydrates only once a day while I eat smaller portions of all food groups. We have not eaten fried foods for about five years". Woman, 56 years, 2 children

In this group, some dietary changes were reported due to health problems such as cholesterol and obesity. As a result, high-fat diets have been reduced.

"I generally eat fish and poultry because I have cholesterol problems, so I try to control my diet. I love colorful dishes with vegetables; this is part of my everyday life". Woman, 56 years, single

"I struggled for three years to reduce my weight, and so I try to have healthier dietary habits such as reducing salt and fat". *Woman, 50 years, single*

Different attitudes were observed between the groups regarding their concerns about their amount of salt intake in diet. Younger women, due to the influence of social media, have on many occasions declared to have ceased the consumption of industrialized soups due to the salt levels in these products, while women over 35 years old knew about the harmful effects of salt, both in sweet and salty recipes, and tried to restrict to partake less frequently in consuming salty products. Women aged over 50 years old mentioned using natural flavoring such as garlic, onion, and herbs in larger quantities to reduce salt intake because of dietary salt restrictions.

"I started to use onions and garlic with more frequency to reduce salt intake". *Woman, 59 years*

"I seasoned foods with lemon juice to reduce salt intake". *Woman, 56 years*

All groups understood the relationship between salt and sodium. For consumers, the sodium content is associated with the amount of salt present in foods. The higher the salt levels, the higher the sodium content, as shown in the statements below:

"I compared the sodium levels of two sausages because I have noticed that the sausage I usually buy was too salty". *Woman, 33 years*

Sausages have been considered an item indispensable in almost all family meals, because it allows for a large variety of recipes. It was observed that the frequency of consumption was an important factor to distinguish one consumer group from another. Younger families with small children try to eat these products only on weekends. Families with school-age children, teens, or even young adults use sausages both to prepare quick meals during the week and snacks on weekends.

"Sausage can be used in pasta sauce, with mashed potato, hot dogs, and even as a snack to accompany beer. I also like to eat it as a snack, without anything added; sometimes, I open the refrigerator and take a sausage to eat". *Woman, 50 years*

Some consumers were concerned about the ingredients used in the preparation of processed meat products, and the greatest concern is related to heterogeneous products, such as burgers, which are considered fatty.

"I'd rather eat sausage than hamburger, because fat is visible in hamburgers but it is not in sausages. If I see fat, I will think about that". *Woman, 40 years*

With respect to the brands, although consumers have reported several sausage brands, they declared to stick to two or three, which are selected according to their experience reinforcing the importance of eating habits. Few consumers mentioned reading the label information in this type of product. The expected sausage characteristics based on the experience are shown in Table 4.

Table 4. Expected characteristics of sausages at the time of purchase, preparation and consumption

Expected characteristic	Description
Appearance in the package	Red color
	Rounded shape
Appearance during cooking	No color fading during cooking
	Keep format during cooking
	Little swelling during cooking
	Smooth appearance after cooking
	Light color inside
Flavor	Salt in the right amount
	Well-seasoned
	Mild flavor
	Meat aftertaste
Texture	Tender
	Smooth texture
	Firm during bite
	Crunchy
	Without particles during chewing
	Consistent
	Compact
Versatility	Good to prepare different dishes
	Allow freezing
	Allow eating uncooked

Exploring the Brazilian Consumer's Perception about Sodium ... 107

Table 5. Sensory attributes considered important by consumers to describe Frankfurt type sausage

Characteristics	Sensory attributes important for sausages
Appearance	Red color
Aroma	Seasoning/smoky aroma
Flavor	Peppery/salty/seasoning flavor (mild/well-seasoned)
Texture	Resistance to bite/easy chewing/tenderness/juiciness/smooth chewing

3.2. Results of Sensory Terms Elicited by Consumers in the Tasting Session

With respect to the appearance of the products, only the descriptors associated with a red color and variations in the hue were mentioned by consumers. Regarding aroma, a spicy or smoky aroma were elicited by participants. Seven flavor descriptors were initially elicited (peppery, salty, seasoning, sausage, sweet, barbecue and sour); however, they did not identify the type of seasoning during the aroma nor during the flavor assessment. The consumers used nine texture descriptors (elastic, rubbery, spongy, consistent, firm, hard, tender, hard peel and juicy) to describe the products. After tasting, some associations among the descriptors were made during the focus group discussion. Some descriptors were replaced by a unique term, e.g., elastic, rubbery, spongy and a consistency that was associated with easy chewing; hard to peel was replaced by resistance to bite; firm, hard and tender were replaced by tenderness. The focus group sessions enabled the identification of 12 descriptors to be used in the subsequent quantitative research, and the final list of sensory attributes is presented in Table 5.

Although the attribute of smooth chewing that describes the presence of particles in the mouth during chewing was not perceived during the tasting session, it was included in the final list because it was mentioned by consumers during the focus group discussions as being an important characteristic to differentiate the quality of the sausage available in the market.

The attributes were similar to those generated by frankfurter consumers to characterize reduced sodium frankfurters using check-all-that-apply questions (Yotsuyamagi, et al. (2015). Their results showed that consumers considered four attributes related to appearance (beautiful, light color, pink and unattractive), nine attributes related to flavor (salty, seasoning, spicy and strange, and their intensities) and three attributes related to texture (juicy, dry and firm). The best before date (expiration date), sodium levels and nutritional information on the label were highly mentioned in the focus group sessions; therefore, they were also included in the list presented in Table 5. Thus, 15 descriptors were analyzed in the quantitative research.

3.3. Quantitative Research

3.3.1. Kano Method

Table 6 shows the classification of the 15 quality attributes according to consumer perception.

From the 15 evaluated quality attributes, only the descriptors of long expiration date and tenderness were considered attractive by consumers. Smooth texture and without particles during chewing were considered proportional. All descriptors related to nutritional information were considered indifferent attributes, as well as the descriptors associated with a specific flavor (peppery, smoky, seasonings, etc.).

Data from the quantitative questionnaire were evaluated for the better/worse scores as proposed by the Kano methodology in order to identify the characteristics that could generate higher or lower consumer satisfaction if they are changed/modified. Figure 1 shows the results:

Exploring the Brazilian Consumer's Perception about Sodium ... 109

Table 6. Frequency of answers and classification of the quality attributes according to the Kano methodology

Quality attributes	Frequency of answers						Kano Classifi-cation
	A	P	E	R	Q	I	
Red color	24	1	10	2	1	82	Indifferent
Longer Expiration date	59	3	26	0	4	28	Attractive
Information on reduced sodium on the label	47	3	5	0	0	65	Indifferent
Nutritional information	25	5	18	0	3	69	Indifferent
Seasoning aroma	19	4	7	0	1	89	Indifferent
Smoky aroma	16	2	1	6	1	94	Indifferent
Resistance to bite	6	6	2	11	2	93	Indifferent
Tenderness	33	11	19	1	0	56	Attractive
Juiciness	23	8	22	0	0	67	Indifferent
Well-seasoned	33	4	20	0	0	63	Indifferent
Mild seasoning	6	9	9	1	0	95	Indifferent
Salty	3	0	1	10	3	103	Indifferent
Peppery	8	0	2	12	0	98	Indifferent
Smooth chewing (without particles)	29	6	32	0	1	52	Expected
Easy chewing and swallowing	14	11	28	2	0	65	Indifferent

A – Attractive; P – Proportional; E – Expected; I – Indifferent; R – Reverse; Q – Questionable.

The characteristics that generate *better* scores (e.g., the greater the better score, the higher the consumers' satisfaction) were long expiration date, smooth texture without the presence of grittiness when chewing, tenderness and well-seasoned. The sodium level highlighted on labels also generated *better* scores, although its absence did not promote displeasure, indicating that, despite not being considered an attractive descriptor, it can generate satisfaction when present, and its absence does not impact on the product's quality, according to consumers. This characteristic can lead to product differentiation at the time of purchase.

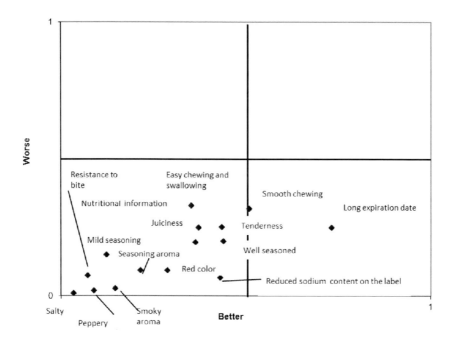

Figure 1. Positioning of the quality attributes according to *better* and *worse* scores.

On the other hand, the attributes that generated *worse* scores, which when absent or reduced will cause consumer dissatisfaction, were easy swallowing and chewing and smoothness when chewing, followed by tenderness and juiciness.

The quality attributes associated with aroma (smoky and seasoned) and flavor (salty, spicy and mild seasoning) showed small *better* and *worse* scores, indicating little influence on the product's quality perception.

3.3.2. Degree of Knowledge and Attitudes towards Salt Intake

Figure 2 shows the knowledge of consumers regarding the harmful effects of salt intake and the daily intake recommended by the World Health Organization.

Exploring the Brazilian Consumer's Perception about Sodium ... 111

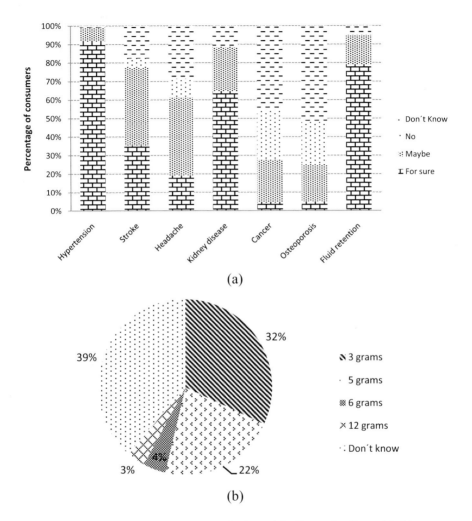

Figure 2. Knowledge of consumers about the harmful effects of sodium chloride (a) and the recommended intake levels (b).

Consumers have proven to know the harmful effects caused by excess sodium chloride, especially hypertension (90% knowledge), fluid retention (79%) and kidney diseases (65%), and ignored the impact of salt on diseases such as cancer and osteoporosis (45% and 52%, respectively). However, only 22% of the consumers recognized the recommended daily amount of sodium chloride (5g), while about 40% of them ignored this dosage.

Table 7. Consumer attitudes about healthy eating habits and sodium chloride intake

	Disagreement (%)	Indifference (%)	Agreement (%)
Healthy eating means eating everything and balanced	14[a]	6[a]	80[b]
Healthiness today is relative then I try eat all foods without excess	18[c]	8[a]	74[b]
I have healthy habits and restrict my diet to natural products without preservatives	56[c]	13[a]	30[b]
I try to restrict as much as possible the consumption of processed products at home	59[c]	9[a]	32[b]
It is important to follow a low-salt diet	3[a]	3[a]	93[b]
A low-salt diet prevents fluid accumulation in my body	5[a]	10[a]	85[b]
A low-salt diet prevents edemas	9[a]	15[a]	76[b]
To make a low-salt diet helps me to breathe more easily	27[a]	41[a]	32[a]
When I make a low-salt diet, I feel better.	10[a]	27[b]	63[c]
A low-salt diet keeps my heart healthy.	3[a]	13[b]	84[c]
My family advised me to follow a low-salt diet.	25[a]	23[a]	52[b]
I like foods with little salt.	30[b]	8[a]	62[c]
I choose low-salt food in restaurants.	30[a]	18[a]	52[b]
I go to restaurants serving low salt foods	44[b]	25[a]	31[ab]
I choose foods with little salt in the supermarket	39[b]	17[a]	45[b]
I want to change my diet in relation to salt intake	23[a]	18[a]	59[b]
I worry about the amount of salt in my diet	16[a]	12[a]	72[b]
My health will improve if I decrease the amount of salt in my diet.	15[a]	21[a]	64[b]
The manufacturer should be more concerned to reduce the salt levels in food.	4[a]	4[a]	92[b]
High amounts of salt are not suitable for children	3[a]	7[a]	91[b]
My family consumes low salt foods	31[b]	15[a]	54[c]
Salt shall be on the table to be used whenever necessary	66[c]	11[a]	23[b]

Exploring the Brazilian Consumer's Perception about Sodium ... 113

Consumers' attitudes about healthy eating and sodium chloride intake habits are shown in Table 7. The majority of the consumers (over 70%) mentioned that healthy eating means eating all foods in a balanced way, without excesses. About 60% of the consumers reported they do not avoid consuming processed foods. Although consumers believe that lower salt intake can prevent health problems such as fluid retention, edema, and heart disease, there was no consensus on their own initiatives to reduce salt intake. About 30 – 40% of the consumers disagreed with the statements: *I like foods with little salt; I choose foods with little salt in restaurants; I choose foods with little salt in supermarkets;* and *I go to restaurants serving low salt foods.*

4. DISCUSSION

Excessive intake of calories was associated with weight gain or fat gain (Fernandes, Oliveira, Rodrigues, Fiates, & Proença, 2015). Heiman and Lowengart (2014) also noted that calorie information affects the perception and the choice process among women, but affects only the perception in men. Of all the variables surveyed on the present study, calorie reduction was barely mentioned, indicating that this appeal should be indirectly less valued when compared to fat reduction. For consumers, processed meat products are considered fatty, especially burgers, because fat is easily perceived (Viana, Silva, & Trindade, 2014). Healthiness was associated with being natural, containing whole ingredients, being rich in fiber, vitamins and minerals and less related to the amount of additives and sodium.

Consumers' awareness about healthy eating habits increases with age (Drichoutis, Lazaridis, & Gaya, 2005; Kenten, Boulay, & Rowe, 2013; Parmenter, Waller, & Wardle, 2000) due to the increase of health problems like obesity and cardiovascular diseases like hypertension (Goldstein & Leshem, 2014). In Brazil (VIGITELBRASIL, 2014); hypertension affects 22.7% of the adult population, and is more common in women (25.4%)

than men (19.5%). These reports were in agreement with the consumers' testimonies in the present study.

UK consumers were unware of the advised salt guidelines, but they were aware of the link between salt intake and health associations such as high blood pressure, water retention and heart disease (Kenten, et al., 2013). Similar results were observed by Newson, et al. (2013) using an online survey on salt intake in eight countries including Brazil. Grimes et al. (2009), and Kim, Lopetcharat, Gerard, and Drake (2012) carried out a study in the metropolitan area of Melbourne, Australia and in the US, respectively, and found very similar values to those observed in the present study regarding the degree of knowledge about the harmful effects concerning sodium chloride intake.

Kim, Lopetcharat, Gerard, and Drake (2012) noticed that although the sodium content on the food label was considered an expected characteristic, it did not influence dissatisfaction and did not necessarily increase consumer product satisfaction. In this study, although sodium content was considered an indifferent characteristic, claims involving its content should contribute to consumer satisfaction.

Habit showed to be an important factor in the consumer's experience as mentioned by Riet, Sijtsema, Degevos, and Bruijn (2011). Other items as "tastes good" and "good value for money" could also explain the food choice according to Carrillo, Varela, Salvador, and Fiszman (2011).

Kock, Zandstra, Sayed, and Wentzel-Vilijoen (2016) observed that 65% of the interviewed South African consumers mentioned not checking labels for salt content at the moment of purchase. The vast majority (92%) of the consumers reported that manufacturers should take more initiatives on sodium reduction. These results corroborate with the results of the qualitative research and are in line with the results reported by Newson, et al. (2013).

CONCLUSION

The results showed that sausage is a food widely consumed by Brazilian consumers, with a strong appeal for convenience during meal preparation and consumption, which pleases the whole family regardless of age. Consumers of this category of products are not concerned about the nutritional information and sodium levels, since the consumers' experience prevails to select the product. However, the sodium levels highlighted on the label was a characteristic that, despite being considered indifferent in the overall assessment, generated great *better* scores, indicating that it could aggregate value to the consumer.

The attribute tenderness was considered an attractive characteristic, which effectively generates consumer satisfaction when present, but its absence does not generate rejection. The homogeneous texture during chewing was considered an expected characteristic. These expected characteristics could be related to the type of raw material used in this product. Brazilian frankfurter generally contains mechanically deboned separated meat that exhibits low shear force and very small bone particles.

Consumers of this product category know the harmful effects of a diet rich in sodium chloride, but are unaware of the recommended levels, thus evidencing the need for continuous actions targeted to education and awareness of the population on this subject.

ACKNOWLEDGMENTS

The authors would like to thank CNPq - National Council for Scientific and Technological Development for the financial support, and LAFISE/ITAL (Food Technology Institute) for recruitment and support to the focus group sessions.

REFERENCES

ABIA. (2013). *Associação Brasileira das Indústrias de Alimentos* [*Brazilian Association of Food Industries*]. http://www.abia.org.br /vs/vs_conteudo.aspx?id=55. Acesso em 06/05/14.

Barcellos, M. D. d., Kügler, J. O., Grunert, K. G., Wezemael, L. V., Pérez-Cueto, F. J. A., Ueland, O., & Verbeke, W. (2010). European consumer's acceptance of beef processing technologies: A focus group study. *Innovative Food Science and Emerging Technologies, 11.*

Cappucio, F. P., Kalaitzidis, R., Duneclift, S., & Eastwood, J. B. (2000). Unravelling the links between calcium excretion, salt intake, hypertension, kidney stones and bone metabolism. *Journal of Nephrology, 13*, 169-177.

Carrillo, E., Varela, P., Salvador, A., & Fiszman, S. (2011). Main factors underlying consumers' food choice: a first step for the understanding of attitudes toward "healthy eating". *Journal of Sensory Studies, 26*, 85-95.

D'Elia, L., Rossi, G., Ippolito, R., Cappuccio, F. P., & Strazzullo, P. (2012). Habitual salt intake and risk of gastric cancer: A meta-analysis of prospective studies. *Clinical Nutrition, 31*, 489-498.

Desmond, E. (2006). Reducing salt: A challenge for the meat industry. *Meat Science, 74*, 188-196.

Despain, D. (2014). Easy steps to less salt. *Food Technology, 1*, 48-50, 52-59.

Dötsch, M., Busch, J., Batenburg, M., Liem, G., Tareilus, E., Meller, R., & Meijer, G. (2009). Strategies to reduce sodium consumption: A food industry perspective. *Critical Reviews in Food Science and Nutrition, 49*, 841-851.

Dötsch, M., Busch, J., Batenburg, M., Liem, G., Tareilus, E., Mueller, R., & Meijer, G. (2009). Strategies to reduce sodium consumption: A food perspective. *Critical Reviews in Food Science and Nutrition, 49*, 841-851.

Exploring the Brazilian Consumer's Perception about Sodium ... 117

Drichoutis, A. C., Lazaridis, P., & Gaya, R. M. (2005). Nutrition knowledge and consumer use of nutritional food labels. *European Review of Agricultural Economics, 32*, 93-118.

Fernandes, A. C., Oliveira, R. C. d., Rodrigues, V. M., Fiates, G. M. R., & Proença, R. P. d. C. (2015). Perceptions of university students regarding calories, food healthiness and the importance of calorie information in menu labelling. *Appetite, 91*, 173-178.

Frisoli, T. M., Schmieder, R. E., Grodzicki, T., & Messerli, F. H. (2012). Salt and Hypertension: Is Salt Dietary Reduction Worth the Effort? *The American Journal of Medicine, 125*, 433-439.

Goldstein, P., & Leshem, M. (2014). Dietary sodium, added salt, and serum sodium associations with growth and depression in the U.S. general population. *Appetite, 79*, 83-90.

Grimes, C. A., Riddell, L. J., & Nowson, C. A. (2009). Consumer knowledge and attitudes to salt intake and labelled salt information. *Appetite, 53*, 189-194.

Guerrero, L., Guàrdia, M. D., Xicola, J., Verbeke, W., Vanhonacker, F., Zakowska-Biemans, S., Sajdakowska, M., Sulmont-Rossé, C., Issanchou, S., Contel, M., Scalvedi, M. L., Granli, B. S., & Hersleth, M. (2009). Consumer-driven definition of traditional food products and innovation in traditional foods. A qualitative cross-cultural study. *Appetite, 52*, 345-354.

Gutman, J. (1982). A means-end chain model based on consumer categorization processes. *Journal of Marketing, 46*, 60-72.

He, F. J., & MacGregor, G. A. (2003). How far should salt intake be reduced? *Hypertension, 42*, 1093-1099.

He, F. J., & MacGregor, G. A. (2009). A comprehensive review on salt and health and current experience of worldwide salt reduction programmes. *Journal of Human Hypertension, 23*, 363-384.

Heiman, A., & Lowengart, O. (2014). Calorie information effects on consumers' food choices: Sources of observed gender heterogeneity. *Journal of Business Research, 67*, 964-973.

Horita, C., Messias, V., Morgano, M., Hayakawa, F., & Pollonio, M. (2014). Textural, microstructural and sensory properties of reduced

118 *Maria T. E. L. Galvão, Rosires Deliza and Marise A. R. Pollonio*

sodium frankfurter sausages containing mechanically deboned poultry meat and blends of chloride salts. *Food Research International, 66,* 29-35.

Horita, C., Morgano, M., Celeghini, R., & Pollonio, M. (2011). Physico-chemical and sensory properties of reduced-fat mortadella prepared with blends of calcium, magnesium and potassium chloride as partial substitutes for sodium chloride. *Meat Science, 89,* 426-433.

Kano, N., Seraku, F. T., & Tsuji, S. (1984). Attractive quality and must-be quality, Hinshitsu. *The Journal of the Japanese Society for Quality Control, 4.*

Kenten, C., Boulay, A., & Rowe, G. (2013). Salt. UK consumers' perceptions and consumption patterns. *Appetite, 70,* 104-111.

Kim, M. K., Lopetcharat, K., Gerard, P. D., & Drake, M. A. (2012). Consumer awareness of salt and sodium reduction and sodium labelling. *Journal of Food Science, 77,* S307-S313.

Knight, P., & Parsons, N. (1988). Action of NaCl and polyphosphates in meat processing: responses of myofibrils to concentrated salt solutions. *Meat Science, 24,* 275-300.

Kock, H. L. D., Zandstra, E. H., Sayed, N., & Wentzel-Vilijoen, E. (2016). Liking, salt taste perception and use of table salt when consuming reduced-salt chicken stew in light of South Africa's new salt regulations. *Appetite, 96,* 383-390.

Krueger, R. A., & Casey, M. A. (2009). *Focus Group - a practical guide for applied research.* 4th Edition. SAGE Publications, USA., 217p.

Kuo, W. Y., & Lee, Y. (2014). Effect of food matrix on saltiness perception - Implications for sodium reduction. *Comprehensive Reviews in Food Science and Food Safety, 13,* 906-923.

McGough, M. M., Sato, T., Rankin, S. A., & Sindelar, J. J. (2012). Reducing sodium levels in frankfurters using a natural flavor enhancer. *Meat Science, 91,* 185-194.

Moskowitz, H. R., Beckley, J. H., & Resurreccion, A. V. A. (2006). *Sensory and consumer research in food product design and development.* IFT Press, Blackwell Publishing, 358p.

Exploring the Brazilian Consumer's Perception about Sodium ... 119

Myers, J. H., & Shocker, A. D. (1981). The nature of product-related attributes. In J.N. Sheth. *Research in Marketing*, 211-236.

Neal, B. (2014). Dietary Salt Is a Public Health Hazard That Requires Vigorous Attack. *Canadian Journal of Cardiology, 30*, 502-506.

Newson, R. S., Elmadfa, I., Biro, G., Cheng, Y., Prakash, V., Rust, P., Barna, M., Lion, R., Meijer, G. W., Neufingerl, N., Szabolcs, I., van Zweden, R., Yang, Y., & Feunekes, G. I. J. (2013). Barriers for progress in salt reduction in the general population. An international study. *Appetite, 71*, 22-31.

Parmenter, K., Waller, J., & Wardle, J. (2000). Demographic variation in nutrition knowledge in England. *Health Education Research, 15*, 163-174.

Riet, J. v. t., Sijtsema, S. j., Degevos, H., & Bruijin, G.-J. d. (2011). The importance of habits in eating behaviour. An overview and recommendations for future research. *Appetite, 57*, 585-596.

Ruusunen, M., & Puolanne, E. (2005). Reducing sodium intake from meat products. *Meat Science, 70*, 531-541.

Sauerwein, E., Bailom, F., Matzler, K., & Hinterhuber, H. H. (1996). The kano model: How to delight your customers. *International Working Seminar on Production Economics, Innsbruk/Igls/Austria, 1*, 313-327.

Shim, S.-M., Seo, S. H., Lee, Y., Moon, G.-I., Kim, M.-S., & Park, H.-H. (2011). Consumer's knowledge and safety perceptions of food additives: Evaluation on the effectiveness of transmitting information on preservatives. *Food Control, 22*, 1054-1060.

Svetkey, L. P., Sacks, F. M., Obarzanek, E., Wollmer, W. M., Appel, L. J., LIn, P., Karanja, N. M., Harsha, D. W., Bray, G. A., Aickin, M., Proschan, M. A., Windhauser, M. M., Swain, J. F., McCarron, P. B., Rhodes, D. G., & Lwas, R. L. (1999). The DASH diet, sodium intake and lood pressure trial (DASH-Sodium) Rationale and design. *Journal of the American Dietetic Association, 99*.

Tobin, B. D., O'Sullivan, M. G., Hamill, R. M., & Kerry, J. P. (2012). Effect of varying salt and fat levels on the sensory and physiochemical quality of frankfurters. *Meat Science, 92*, 659-666.

120 *Maria T. E. L. Galvão, Rosires Deliza and Marise A. R. Pollonio*

Toldrà, F., & Reig, M. (2011). Innovations for healthier processed meats. *Trends in Food Science and Technology, 22*, 517-522.

van Kleef, E., van Trijp, H. C. M., & Luning, P. (2005). Consumer research in the early stages of new product development: a critical review of methods and techniques. *Food Quality and Preference, 16*, 181-201.

Verbeke, W., Pérez-Cueto, F. J. A., Barcellos, M. D. d., Krystallis, A., & Grunert, K. G. (2010). European citizen and consumer attitudes and preferences regarding beef and pork. *Meat Science, 84*, 284-292.

Viana, M. M., Silva, V. L. d. S., & Trindade, M. A. (2014). Consumers' perception of beef burgers with different healthy attributes. *LWT - Food Science and Technology, 59*, 1227-1232.

Vigitel Brasil. (2014). *Vigilância de fatores de risco e proteção para doenças crônicas por inquérito telefônico.* Ministério da Saúde, Secretaria de Vigilância em Saúde, Departamento de Vigilância de Doenças e Agravos não Transmissíveis e Promoção da Saúde. – Brasília: Ministério da Saúde, 152p [Surveillance of risk factors and protection for chronic diseases by telephone survey. Ministry of Health, Secretariat of Health Surveillance, Department of Surveillance of Diseases and Noncommunicable Diseases and Health Promotion. - Brasília: Ministry of Health, 152p].

Weiss, J., Gibbs, M., Schuh, V., & Salminen, H. (2010). Advances in ingredient and processing systems for meat and meat products. *Meat Science, 76*, 196-213.

WHO. (2015). Healthy diet. Fact Sheet n. 394 In: *World Health Organization.* http://www.who.int/mediacentre/factsheets/fs394/en/. Access in 02/24/2015.

Yotsuyamagi, S. E., Contreras-Castillo, C. J., Haguiwara, M. M. H., Cipolli, K. M. V. A. B., Lemos, A. L. S. C., Morgano, M., & Yamanda, E. A. (2015). Tecnhological, sensory and microbiological impacts of sodium reduction in frankfurters. *Meat Science*, http://dx. doi.org/10.1016/j.meatsci.2015.1012.1016.

Zandstra, E. H., Lion, R., & Newson, R. S. (2016). Salt reduction: Moving from consumer awareness to action. *Food Quality and Preference, 48*, 376-381.

In: Beef: Production and Management Practices ISBN: 978-1-53613-254-0
Editor: Nelson Roberto Furquim © 2018 Nova Science Publishers, Inc.

Chapter 5

THE RELATIONSHIP OF FATTY ACID COMPOSITION, CHOLESTEROL CONTENT, NUTRITIONAL AND ENZYMATIC INDICES WITH THE INTRAMUSCULAR FAT CONTENT OF BEEF

Pilar T. Garcia[1,3,], Phd, Nestor N. Latimori[2], Ana M. Sancho[1] and Jorge J. Casal[3]*

[1]Instituto Tecnologia Alimentos,
CIA, INTA Castelar (1712), Pcia Bs As, Argentina
[2]Estacion Experimental INTA Marcos Juarez, Cordoba, Argentina
[3]Facultad Agronomia y Ciencias Agroalimentarias,
Universidad de Moron, Pcia Bs As. Argentina

---

[*] Corresponding Author Email: garcia.pilar@inta.gob.ar.

124 *Pilar T. Garcia, Nestor N. Latimori, Ana M. Sancho et al.*

ABSTRACT

Nutritional value of beef is related to its fatty acid composition and the amount of intramuscular fat (IMF). The aim of this study was to find out the relationships between the fatty acid (FA) composition, cholesterol (CHO) content and selected nutritional and enzymatic indices (SI) with the IMF content of beef from three genotypes Aberdeen Angus (AA), Charolaise x Angus (CHA) and Argentine Holstein (HA) under four production systems, pasture exclusive (P), pasture and 0.7% corn grain (Cn0.7%, pasture and 1% corn grain (Cn1%) and 85% corn (Cn85%). The FA profile, determined by diet and genotype, affects the nutritional and enzymatic indices and its level of correlation with IMF content. Correlations of IMF content with polyunsaturated fatty acids (PUFAs) were generally negative, whereas, with monounsaturated fatty acids (MUFAs) were positive and very low. Correlations of IMF content with PUFA/SFA and thrombogenic index (TI) were generally negatives and positives with atherogenic index (AI). Correlations with enzymatic indices were variables. Fatty acid profile had shown a wide variability depending on factors such as genotype and diet. The overall intramuscular fat content had variable effect on relative fatty acid composition and, consequently, on nutritional and enzymatic indices values. The intramuscular fat content had impacted on relative fatty acid composition of neutral and phospholipids with implications on the nutritional quality of beef. By identifying the natural variations in the FA profile of beef, producers could significantly improve beef nutritional quality with a combination of diet and genetic selection.

Keywords: beef intramuscular fat, fatty acids composition, cholesterol, enzymatic and nutritional indices

1. INTRODUCTION

The nutritional value of beef meat is related to its fatty acid (FA) composition and the amount of intramuscular fat (IMF). The amount of intramuscular fat or marbling deposited in *Longissimus* muscle is a major determinant of carcass value and predictor of meat quality. Beef IMF comprises over 20 individual fatty acids, however, six mayor FAs contribute over 92% of total FAs. These FAs are myristic (C14:0), palmitic

(C16:0), palmitoleic (C16:1 n-7), stearic (C18:0), oleic (C18:1 n-9) and linoleic (C18:2 n-6).

Ruminant meats are of particular consumer concern because of the high level of saturated fatty acids (SFAs) when compared to pork or poultry. However, ruminant meat holds great potential in becoming a functional food due to the presence of some monounsaturated fatty acids (MUFAs), polyunsaturated fatty acids (PUFAs) in particular long-chain n-3 LC-PUFAs, and the conjugated linoleic acid (CLA) isomers, which have favorable effects on human health (Lee et al., 2005; Daley et al., 2010).

The FAs composition of beef lipids is determined by genetic factors such as breed, gender and genotype, and environmental factors of which diet is by far the most important one. Differences in the concentrations of fatty acids, in the ratio n-6/n-3 fatty acids and the amounts of conjugated isomers of linoleic acid (CLA) in beef lipids have been detected due to animal diet differences (Garcia, 2012; Garcia et al., 2017) Grasses are rich in linolenic acid (n-3 fatty acid) and grains generally rich in linoleic acid (n-6 fatty acid). The amounts of conjugated isomers of linoleic acid (CLA) were also higher in grass beef lipids compared with the grain ones (Garcia et al., 2008; Garcia et al., 2012). The forage types, crop variety, cutting, season, year, etc., affects the fatty acid composition of forage crops for grazing forage beef production (Garcia et al., 2015). Breed also affects the fatty acid composition of muscle lipids. Malau-Aduli et al., (1998) found differences between Jersey and Limousine cattle in fatty acid composition of muscle phospholipids. Laborde et al., (2001) found differences in fatty acid composition between Simmental and Red Angus although suggested that the biological and practical significance needs to be demonstrated.

There are three desaturases in animal tissues $\Delta 5$, $\Delta 6$ and $\Delta 9$. Of these only $\Delta 9$ desaturase acts upon SFAs to convert them to their respective MUFAs. The most abundant FA in beef is oleic acid produced by the desaturation of stearic acid. The expression of stearoyl-CoA desaturase (SCD) is associated with adipocyte hypertrophy in a number of species. C18:2n-6 and C18:3n-3 serve as the precursor molecules from which the LC-PUFAs belonging to the n-3 and n-6 fatty acid family can be synthesized through a series of elongation and desaturation reactions. All the reactions are catalyzed by an enzymatic system consisting in fatty acyl-CoA synthetases Δ-6 and Δ-5 desaturases and respective elongases. These two fatty acid families not only share these enzymes, but also compete for the same enzymes. A key function of C18:3 n-3 is a substrate for the synthesis of longer-chain omega 3 fatty acid EPA (C20:5 n-3) and DHA (C22:6 n-3), which play an important role in the regulation of inflammatory immune reactions and blood pressure, brain development, cognitive function, etc.

The cis-9, trans-11 isomer (CLA) can be produced in the rumen via biohydrogenation and in the tissues from a reaction catalyzed by stearoyl-CoA desaturase (SCD). Therefore, increasing the SCD expression and/or activity may be one way to enhance CLA concentration in tissues. Also, SCD enzyme is related to the biosynthesis of the other health beneficial fatty acids such as MUFAs (Choi et al., 2002). The key enzymes involved in SFAs production are fatty acid synthase (FAS) and acetyl-CoA-carboxylase α (AACα). The biosynthesis of MUFA is catalyzed by sterol-CoA desaturase (SDC) and the key enzymes in biosynthesis of PUFA are Δ6-desaturase (Δ6d) and Δ5-desaturase (Δ5d).

The aim of this work was to study the relationship of several fatty acids, cholesterol content, and nutritional and enzymatic indices with the intramuscular fat content of steers from four feeding systems and three different genotypes. The feeding systems described here, represent productive models widespread in Argentina Pampeana region, ranging from pasture exclusive diet, to the feedlot system. The British purebred or their crossbreds, which are the main components of the meat consumed domestically, were represented by Aberdeen Angus (AA) steers. The crossbred Charolaise x Aberdeen Angus (CHA) steers represents the most advisable crossbred for production of heavy steers for export markets, and Argentine Holstein (HA) steers are potentially adequate for commercial purpose.

2. MATERIALS AND METHODS

2.1. Animal Management of the Animals and Their Diets

A total of 144 Aberdeen Angus (AA), Charolaise x Aberdeen Angus (CHA) and Argentine Holstein (HA) steers were allotted from 5-7 months old to slaughter weight to the four following treatments:

Pasture (P): steers grazed on pasture exclusively (mainly alfalfa and festuca).

Cn0.7%: steers fed with a similar pasture but supplemented daily with cracked corn grain (0.7% of live weight).

Cn1%: steers fed with a similar pasture but supplemented daily with cracked corn grain (1% of live-weight).

Cn85%: steers fed with a diet based in corn grain (85%), alfalfa hay (8.6%), soy bean meal (8.4%) and minerals and vitamins (0.5%).

The forage in P, Cn0.7% and Cn1% treatments were provided by a mixed pasture of alfalfa (*Medicago sativa*) and tall fescue (*Festuca arundinacea Schreb*) grazed under a rotational system of six paddocks with seven days of permanency in each one. Twelve steers for treatment were assigned and they were conventionally slaughtered in a conventional abattoir at a similar degree of finishing estimated by visual evaluation. Data from the animal performance within each breed and treatment group are presented in Tables 1 and 2.

Table 1. Animal data: Initial and final live weights and experimental period (EP)

Genotype	Initial live weight (kg)			Final live weight (kg)			EP (days)		
	AA	CHA	HA	AA	CHA	HO	AA	CHA	HA
N	12	12	12	12	12	12	12	12	12
Pasture	190a	181a	195a	388b	449a	505b	365	382	391
Cn0.7%	188a	171a	194a	408a	467a	512ab	333	378	367
Cn1%	188a	174a	194a	420a	483a	541a	289	356	354
Cn85%	189a	168a	193a	350a	399b	374c	152	192	170

Note: a, b, c means within a column with different letters significantly differ (p < 0.05).

128   Pilar T. Garcia, Nestor N. Latimori, Ana M. Sancho et al.

Table 2. Experimental animals: growth rate (g/d) and EAC (estimated average contribution) of grain to total energy intake

	Growth rate g/d)			EAC
Genotype	AA	CHA	HA	
Pasture	546c	707c	800d	0%
Cn0.7%	675bc	787c	875cd	33.6%
Cn1%	808b	871bc	983bc	44.9%
n85%	1055a	1223a	1093ab	85:8%

Note: a, b, c, d means within a column with different letters significantly differ (p < 0.05).

2.2. Methods of Analysis

After 24 h at 4°C, 12 steaks of 2-25 cm of thickness, from *Longissimus dorsi* muscles at the 11th rib were taken from each group, carefully dissected and used for chemical analysis. All samples were frozen and stored at -20°C until the analysis was performed. Aliquot samples of 10g each, trimmed of external fat, minced carefully, dried and extracted in a Tecator apparatus using hexane as the extraction solvent according to official methods (AOAC, 1992), were used to determine total intramuscular fat (IMF). Aliquot samples of 5g each were extracted using the Folch et al., method (Folch et al., 1957). Fatty acid methyl esters (FAME) were prepared according to the method of Pariza et al., (2001) and measured using a Chrompack CP 900 equipment (Chrompack Inc., Middleburg, The Netherlands) fitted with a flame ionization detector. Fatty acid methyl esters were separated with a fused silica capillary column CP-Sil 88 (100 m x 0.25 mm i.d.) from Chrompack Inc., Middleburg, The Netherlands, with N2 as the carrier gas. The oven temperature was programmed at 70°C for 4 min, increased from 70 to 170°C at a rate of 13°C/min and then increases from to 170° to 200°C at 1°C/min. Individual fatty acids were identified by comparing relative retention times with individual fatty acids standard (PUFA-2 Animal Source. Supelco). Analytic results were expressed as percentages of total fatty acids and mg of fatty acid contribution in 100g of fresh *Longissimus dorsi* muscle. One

The Relationship of Fatty Acid Composition ... 129

aliquot sample of the chloroform extract was saponified with ethanol absolute 50% OHK, extracted and the total cholesterol determined by a colorimetric enzymatic method (BioSystem S.A.).

2.3. Statistical Analyses

Data was subjected to analysis of variance using the GLM (SAS 9.1 SAS Institute, Inc., Cary, NC) statistical software package. A two factorial design was considered: genotype with three levels (AA, CHA and HA) and diet with four levels (Pasture, pasture + 0.7% grain, pasture + 1% grain and 85% grain) and their interactions. In case of a significant treatment effect by F-test, the Tukey`s studentized range (HSD) was used for follow-up comparisons of treatment means. The fatty acid selected were SFAs (C14:0, C16:0 and C18:0), MUFAs (C16:1 n-7 and, C8:1 n-9), n-6 PUFAs (18:2, C20:3 and C20:4) and n-3 PUFAs (C18:3, C20:5, C22:5 and C22:6). The Δ-desaturase index, as an indirect index of stearoyl-CoA desaturase (SCD) activity, was calculated as C18:1/C18:0 (SCD18) or C16:1/C16:0 (SCD16). The elongase index was calculated as the ratio of C18:0 to C16:0, whereas the thioesterase index was calculated as the ratio of C16:0/C14:0 (Zhang et al., 2007). The indices of atherogenic (AI) and thrombogenic (TI) were calculated as proposed by Ubricht & Southgate (1991).

3. RESULTS AND DISCUSSION

3.1. Effects of Diet, Genotype and IMF Content on Beef Lipid Fatty Acids

The fatty acid composition of IMF, expressed as percentages of total lipids, is presented in Table 3. Fatty acid profile shows a wide variability depending on diet and genotype. Changing the FA composition of beef has typically been done by dietary means, but deposition of FAs depends on the pathways and rates of the ruminal PUFAs biohydrogenation, and genetic factors governing the deposition and turnover rates of individual FA (Kramer et al., 2004).

3.1.1. Saturated Fatty Acids

C14:0 acid was affected by genotype (AA = CHA > HA) and diet (Cn85% > P = Cn0.7% = Cn1%). C16:0 was only affected by genotype (AA > CHA > HA) and C18:0 by genotype (AA = CHA > HA) and diet (P > Cn0.7% > Cn1% > Cn85%). Bellizzi et al., (1994) found a very strong relation of C14:0 with coronary heart disease. This is due to the repression of the LDL receptor synthesis as well as direct stimulation of hepatic LDL synthesis. C16:0 was also associated with an increase in cholesterol concentrations (Yu et al., 1995). C18:0 has a neutral effect on total serum cholesterol and no effects on LDL and HDL (Mensing, 2005). From a nutritional point of view, while C18:0 acid is thought to be neutral in influencing plasma cholesterol, C14:0 and C16:0 are significantly associated with coronary health risk, due to their cholesterol raising effect (McAfee et al., 2010).

The relationships between SFAs percentages with IMF content varied in levels of significance and correlation among diets and genotypes. As presented in Table 4, Pasture-fed SFAs were no correlated with the IMF content. C14:0 was negatively correlated with IMF content in AA Cn85% (r = -0.71**), CHA Cn1% (r = -0.65*) and HA Cn1% (r = -0.75**), whereas C16:0 acid percentage was only positively correlated with IMF content in Cn0.7% and Cn1% diets. C18:0 was not correlated with IMF content. Brugiapaglia et al., (2013) report similar results. The increase of SFAs is caused by the decrease in the phospholipids neutral lipids ratio that arises from the increase in the IMF. Indeed, total SFAs are presented in the neutral fraction, whereas total PUFAs are present mainly in the phospholipids fraction.

Table 3. Fatty acid proportions of *Longissimus dorsi* muscle in AA, CHA and HA genotypes, and Pasture, Cn0.7%, Cn1% and Cn85% diets

Fatty acid	Genotype AA CHA HA			Diet Pasture Cn0.7% Cn1% Cn85%				RESM	p value Genotype Diet GXD		
IMF %	3.57	3.80	3.59	2.89b	3.58ab	4.25a	3.90a	1.32	NS	0.001	NS
C14:0	2.45a	2.36a	2.24b	2.24b	2.21b	2.27b	2.67a	0.39	0.034	0.001	NS
C16:0	24.49	23.90	23.09	23.85	23.82	24.26	23.38	1.71	0.001	NS	0.01
C16:1	3.34b	3.30b	4.01a	3.33b	3.52ab	3.63a	3.72a	0.44	0.001	0.001	NS
C18:0	13.23a	13.17a	11.70b	13.53a	13.14ab	12.71b	11.55c	1.22	0.001	0.001	NS
C18:1	34.01b	34.90b	36.61a	33.17b	35.67a	36.50a	36.35a	3.27	0.001	0.001	NS
C18:2	4.10	4.30	4.29	3.55b	3.76b	3.94b	5.65a	0.97	NS	0.001	NS
C20:3	0.44	0.42	0.44	0.44	0.41	0.42	0.46	0.16	NS	NS	NS
C20:4	1.10b	1.16b	1.41a	1.12	1.18	1.23	1.35	0.48	0.0052	NS	NS
C18:3	0.91a	0.93a	0.81b	1.38a	0.99b	0.77c	0.39d	0.20	0.008	0.001	NS
C20:5	0.31	0.31	0.35	0.57a	0.34b	0.25b	0.14c	0.13	NS	0.001	NS
C22:5	0.52	0.50	0.52	0.76a	0.53b	0.47b	0.30c	0.20	NS	0.001	NS
C22:6	0.33	0.29	0.13	0.38	0.23	0.20	0.17	0.13	0.001	0.001	0.001
CLA	0.50	0.55	0.58	0.68	0.63	0.56	0.30	0.09	0.001	0.001	0.01

Note: Significance of main effects (genotype and diet) and their interactions (GXD).

a, b, c, d means within a row with different letters significantly differ. RESM (Root mean square error).

132   *Pilar T. Garcia, Nestor N. Latimori, Ana M. Sancho et al.*

3.1.2. Monounsaturated Fatty Acids

MUFAs C14:1, C16:1 n-7 and C18:1 n-9 were affected by genotype and diet (Table 3). The higher values were for HA compared with AA and CHA. All MUFAs increases as the grain in the diet increases. The increase of MUFAs content of tissue in response to concentrate-rich diets is associated with an increase in stearoyl CoA desaturase gene expression (Daniel et al., 2004). C14:1 and C16:1 were only significantly correlated with IMF content (Table 4) in Cn0.7% and Cn1% diets: AA (r = 0.57*to 0.64*); CHA Cn0.7% (r = 0.57* to 0.92**) and CHA Cn1% (r = -0.75** to-0.69**) and HA Cn1% (r = 0.50-0.58*). The C18:1 correlations with IMF were small and only in CHA and HA Cn0.7% (r = 0.57*). These results were not consistent with data of Indurain et al., (2006) who reported a positive correlation between MUFAs, oleic acid and IMF content in Pirenaica bulls and (Raes et al., (2001) who stated that the proportions of C18:1 diminished with a decrease in IMF in beef.

3.1.3. Polyunsaturated Fatty Acids

The n-6 PUFA C20:3 was not affected by diet or genotype, whereas C20:4 was only affected by genotype (HA > CHA = AA) (Table 3). At similar amounts of C18:2 n-6 any difference in C20:4 n-6 between breeds only could be explained by a genetic difference due to a lesser Δ^5 desaturase activity responsible for the conversion of C20:3n-6 to C20:4n-6 or lower elongase activity as a result of the relative accumulation of C18:3 n-6. The n-3 PUFAs decreased linearly as the amounts of grain in the diet increased (P > Cn0.7% > Cn1% > Cn85%). C18:3n-3 and C22:6n-3 were also affected by genotype (AA > CHA > HA). A significant (p > 0.01) interaction diet x genotype was detected for C22:6 n-3. The hyperlipidemic effects of C20:4 n-6 are counteracted by C20:5 n-3 (Scientific Review Committee, 1990). The biosynthesis of long chain PUFAs from C18:3 n-3 appears to be a minor pathway in several species. In contrast, C20:5 n-3 may be well utilized for synthesis of other long–chain PUFAs such as C22:6 n-3. A feedback control mechanism responsive to the plasma concentrations of C22:6 n-3 may affect processes that regulate its own synthesis, thereby maintaining C22:6 n-3 homeostasis during dietary

The Relationship of Fatty Acid Composition ...   133

changes (Pawlosky et al., 2001). PUFAs and most FAs in this category were inversely correlated with IMF content (Table 4).

Correlations between individual PUFAs and IMF content were generally negative or not significant, C18:2 n-6 (r = - 0.11 to - 0.86**); C18:3n-3 (r = -0.06 to -0.66*); C20:3 n-6 (r = - 0.09 to -0.70**); C20:4 n-6 (r = - 0.13 to -0.85**); C20:5 n-3 (r = - 0.03 to -0.88**); C22:5 n-3 (r = - 0.06 to -0.88**) and C22:6 n-3 (r = -0.01 to -0.67*) according to genotype and diet. Results of this study confirmed that the proportions of C18:2 n-6 decline as IMF content increases. However, the proportion of C18:3 n-3 did not decrease at the same level when IMF increased. (r =- 0.03, p > 0.05). These results indicate that LD muscle with greater IMF tended to have a smaller ratio of C18:2 n-6/C18:3 n-3 or even n-6: n-3 PUFA, which is desirable not only because IMF improves eating quality, but also because reduction of n-6/n-3 in beef benefits human health. The n-6/n-3 ratio is considered as a nutritional index for the healthiness of food for human consumption and it should not exceed a value of 4 in the human diet to prevent the occurrence of cardiovascular diseases (FAO Food and Nutrition, 2010).

CLA percentages were affected by genotype (HA > CHA > AA) and diet (P > Cn0.7% > Cn1% > Cn85%) (Table 3). A significant (p > 0.01) interaction diet x genotype was detected. The concentration of CLA in fat is higher in ruminants fed on leafy grasses than in those fed on stored forages or concentrates. Increased PUFA in the diet may limit ruminal production of CLA and VA and/or may depress stearoyl CoA desaturase expression or activity in lean tissues, which in turn limits CLA formation and accretion in tissues. Increased dietary forage tended to increase C18:0, C18:2 n-6 and C18:3 n-3 suggesting that increased forage may mitigate toxic effects of PUFA on ruminal biohydrogenation thereby increasing the pool of CLA and VA available for CLA formation and accretion in tissues. Rumenic acid distribution is not similar to other PUFAs, as it was independent of fat location. In ruminant muscles, it is known that CLA is mainly associated to the triacylglycerol fraction, which is linked to the fat content of tissues. (Bauchard et al., 2005). No significant correlations of CLA with IMF were detected (Table 4).

Table 4. Pearson correlation coefficients between fatty acid proportions and intramuscular fat content of *Longissimus dorsi* muscle

FA	P AA CHA HA	Cn0.7% AA CHA HA	Cn1. % AA CHA HA	Cn85% AA CHA HA
C14:0	0.18 0.48 0.09	0.38 0.42 0.47	-0.10 -0.65* 0.75**	0.71** 0.32 0.09
C14:1	0.33 0.03 -0.20	0.62* 0.92** 0.06	0.64* -0.69* 0.50	0.44 0.19 -0.38
C16:0	0.44 0.46 -0.17	0.67* 0.68* -0.05	0.58* 0.23 0.72**	0.21 0.30 0.19
C16:1	-0.12 0.27 -0.14	0.57* 0.57* 0.41	0.60* -0.75** 0.58*	-0.41 0.11 -0.51
C18:0	0.08 0.51 0.29	-0.26 -0.37 -0.02	0.02 0.49 0.42	-0.01 0.04 0.29
C18:1	0.03 -0.06 -0.01	0.01 0.57* 0.57*	0.13 -0.20 0.42	-0.27 -0.09 0.14
C18:2	-0.33 -0.40 0.13	-0.61* -0.76** -0.89**	-0.67* -0.34 -0.74**	-0.57* -0.32 -0.11
C18.3	-0.08 -0.02 0.23	-0.31 -0.58* -0.66*	-0.29 0.31 -0.57*	-0.31 -0.06 0.13
CLA	-0.04 -0.36 0.05	0.48 0.16 0.44	0.43 -0.33 0.35	0.24 0.06 -0.28
C20:2	-0.18 -0.44 -0.25	0.47 -0.40 -0.65*	0.03 -0.48 -0.53*	-0.42 0.23 -0.57*
C20:3 n-6	-0.40 -0.36 -0.20	0.09 -0.72** -0.82**	-0.70** -0.42 -0.50	0.57* -0.29 -0.42
C20:4 n-6	-0.57* -0.39 -0.13	-0.76** -0.70** -0.72**	-0.85** -0.13 -0.76**	-0.60* -0.66* -0.62*
C20:5 n-3	-0.46 -0.47 -0.10	-0.46 -0.60* -0.84**	-0.88** -0.76** -0.38	-0.67* 0.06 -0.03
C22:5 n-3	-0.58* -0.46 -0.35	-0.44 -0.03 -0.50	-0.88** -0.41 -0.47	-0.32 -0.28 -0.27
C22:6 n-3	-0.07 -0.40 0.04	-0.18 -0.02 -0.62*	-0.67 * 0.13 0.07	-0.36 -0.36 0.32

Note: *P < 0.05 ** P < 0.01.

3.2. Effects of Diet, Genotype and IMF Content on Nutritional Indices

Diet affected significantly all nutritional indices considered in this study, whereas, genotype has no effects on n-6 PUFAs, IMF%, total PUFAs, PUFA/SFA, TI and thioestearase indices (Table 5).

3.2.1. n-3 and n-6 PUFAs

Total n-3 PUFAs percentages were affected by genotype (AA > CHA > HA) and diet (P > Cn0.7% > Cn1% > Cn85%). However, n-6 PUFAs were only affected by diet (P< Cn0.7% < Cn1% < Cn85%). The n-6/n-3 and C8:2n-6/C18:3n-3 ratios were affected by genotype (HA > AA = CHA) and for diet (Cn85 > Cn1% > Cn0.7%>P). The n-6/n-3 PUFA ratio is an important index to evaluate the nutritional value of fat (FAO 2010). It is interesting to notice that EPA and DHA only are formed in significant amounts if the ratio n-6/n-3 is less than 10:1. IMF% were only affected significantly by diet (P < Cn0.7% < Cn1% < Cn85%).

The correlation of nutritional indices with the IMF content is presented in Table 6. Total n-3 PUFAs was generally negatively correlated with IMF content. The highest values were with Cn0.7% diet (r = -0.93**HA) and in Cn1% (r = -0.80** AA). Total n-6 PUFAs was generally negatively correlated with IMF content. The highest values were for Cn0.7% diet (r = -0.72** AA), (r = -0.77** CHA) and (r = -0.91** HA). The n-6/n-3 ratio was only correlated with IMF content in Cn0.7% (r = -0.62* CHA) and (r = -0.58* HA) and the n-3/n-6 ratio was only correlated in Cn0.7% (r = 0.57*CHA) and (r = 0.58* HA). The C18:2n-6/ C18:3 n-3 was only correlated with IMF in P (r = -0.60* CHA) and in Cn 0.7% (r = -0.60* HA). The C20:4n-6/C20:5 n-3 was not correlated with IMF in any diet or genotype.

3.2.2. Saturated, Monounsaturated and Polyunsaturated Indices

SFAs % were affected by genotype (AA > CHA > HA) and for diet (P > Cn0.7% > Cn1% > Cn85%.); MUFAs % were affected by genotype (HA > AA = CHA) and diet (Cn85 = Cn1% = Cn0.7% < P); PUFAs % were affected only by diet (Cn85 = P > Cn0.7%= Cn1%) (Table 5). The correlation analysis (Table 6) showed that the SFA percentages were only correlated with IMF content in CHA (r = 0.57* P) and in HA (r = 0.77 Cn1%). An increase in IMF normally results in the continuous deposition of neutral lipids (Indurain et al., 2006). Neutral lipids, predominantly triacylglycerol's, are rich in SFAs and MUFAs with approximately 80% of fatty acid C16:0, C18:0 and C18:1 because of the mechanism of the *novo* FA synthesis, whereas polar lipids, predominantly phospholipids, have a much greater PUFA content.MUFAs % were affected and genotype (HA > AA = CHA) and diet (Cn85% = Cn1.0% = Cn0.7% > P). MUFAs % were positively correlated only with IMF% in Cn0.7% (r = 0.64*CHA) and (r = 0.74**HA). PUFAs % remains relatively constant across different beef diets. This is mainly due to the relatively constant proportion of phospholipids in the cell membranes, and increasing deposition of triglycerides in the adipocytes with increasing intramuscular fat content. The relationship PUFAs % with IMF % was negatively significant in Cn0.7% (r = -0.69* AA), (r = -0.67* CHA) and (r = -0.94** HA), in Cn1% (r = -0.76** AA) and (r = -,0.75** HA) and in Cn85 (r = -0.60* HA).

PUFA/SFA ratio was affected significantly by diet being significantly higher in Cn85% compared with the other diets. Our obtained values ranged from 0.20 to 0.24. Nogalski et al., (2014) and Costa et al., (2006) found values ranging from 0.07 and 0.18, depending on body weight, muscle type, slaughter season, and gender of animals. The PUFA/SFA ratio was not correlated with IMF content in any genotype in the P diet, however, in the Cn0.7% was correlated in the three genotypes AA (r = -0.71**); CHA (r = -0.70**) and in HA (r = -0.90**) and in the Cn1% in AA (r = -0.77**) and in HA(r = -0.74**). The PUFA/SFA ratio has frequently been used to interpret FA composition and to evaluate the nutritional value of fat (Alfaia et al., 2006). The MUFA/SFA ratio was affected by diet and genotype being higher in HA and Cn85%.

MUFA/SFA ratio was not correlated with IMF content. C18:2n-6/CLA ratio was affected by genotype (AA > CHA > HA) and diet (Cn85% > Cn1% > Cn0.7% > P); C18:2/CLA ratio was only related with IMF content in Cn0.7% (r = -0.56*HA). C18:3 n-3/CLA was affected by genotype (AA > CHA > HA) and diet (P > Cn0.7% > Cn1% > Cn85%); C18:3/CLA ratio was related with IMF content in Cn0.7% (r =-0.74** HA) and in Cn1% (r = -0.60*HA).

AI index was affected by genotype (AA = CHA > HA) and diet (P = Cn0.7% > Cn1% = Cn85%) and positively correlated with IMF content in P (r = 0.74** AA); Cn1% (r = 0.82** HA) and in Cn85% (r = 0.58* AA). TI index was affected only by diet (Cn85%> P = Cn0.7% = Cn1%) and thioesterase (C16:0/C14:0), responsible for terminating the cycles of FA synthesis and the realizing the newly synthesized FA, was affected only by diet. The TI was negatively correlated with Cn0.7% (r =-0.65* AA), (r = -0.63 CHA) and (r = -0.89** HA); with Cn1% (r = -0.62* AA) and (r = -0.69** HA). The indices of atherogenic (AI) and thrombogenic (TI) take into account the different effects that single FA might have on human health and in particular on the probability of increasing the incidence of pathogenic phenomena, such as atheroma and or/ thrombus formation. The recommended values of the AI are below 0.5 (Ulbricht & Southgate, 1991).

3.3. Effects of Diet, Genotype and IMF Content on Enzymatic Indices

The effects of genotype and diet on enzymatic indices of desaturases (C16:1/C16:0 and C18:1/C18:0), elongase (C18:1/C16:1), and combined effects of desaturases and elongases (C20:3/C18:2, C20:4/C18:2) in n-6 PUFA and elongase (C22:5/C20:5) and combined effects of desaturases and elongases (C20:5/C18:3 and C22:5/C18:3) in n-3 PUFA involved in PUFA metabolism in *Longissimus dorsi* muscle in steers are shown in Table 7.

Table 5. Nutritional indices of *Longissimus dorsi* muscle in AA, CHA and HA genotypes

Item	Genotype AA CHA HA			Diet Pasture Cn0.7% Cn1% Cn85%				RESM	Genotype	Diet	GxD
n-3	2.07a	2.03ab	1.82b	3.09a	2.09b	1.69c	1.01d	0.49	0.05	0.001	NS
n-6	5.97	6.24	6.41	5.48b	5.61b	5.91b	7.82a	1.40	NS	0.001	NS
n-6/n-3	3.84	3.78	4.71	1.08	2.69	3.61	8.32	1.34	0.001	0.001	0.01
C18:2/C18:3	6.36b	6.72b	8.05a	2.59c	3.84bc	5.31b	16.43a	3.59	0.05	0.001	NS
C20:4/C20:5	4.36	5.71	6.79	2.01	3.55	5.28	11.63	3.03	0.01	0.001	0.01
IMF %	3.57	3.80	3.59	2.89b	3.58ab	4.25a	3.90a	1.32	NS	0.001	NS
Cholesterol [1]	45.3a	44.2ab	40.8b	40.3b	45.1b	42.4ab	45.8	2.55	0.001	0.001	NS
SFA %	40.25	39.43	37.03	39.61	39.16	39.24	37.60	2.54	0.001	0.01	0.01
MUFA %	38.93b	39.93b	42.02a	38.11	40.55a	41.25a	41.25a	2.52	0.001	0.001	NS
PUFA %	8.04	8.27	8.22	8.57ab	7.70b	761b	8.83a	1.76	NS	0.01	NS
P/S	0.20	0.21	0.22	0.22ab	0.20b	0.20b	0.24a	0.05	NS	0.01	NS
MUFA/SFA	0.96b	1.01b	1.13a	0.96b	1.04a	1.05a	1.10a	0.23	0.001	0.001	NS
AI	0.49a	0.47a	0.41b	0.49a	0.46ab	0..44b	0.44b	0.03	0.001	0.001	NS
TI	7.93	8.21	7.62	7.35b	7.34b	7.65b	9.35a	2.96	NS	0.001	NS
Thioesterase	10.25	10.40	10.60	10.78a	10.37a	10.75a	8.98b	1.93	NS	0.001	NS

Note: Significance of main effects (genotype and diet) and their interactions (GxD). [1] mg/100g abcd means within a row with different letters significantly differ. RESM (Root mean square error).

Table 6. Pearson correlation coefficients between some nutritional indices and intramuscular fat content of *Longissimus dorsi* muscle

FA	Pasture AA CHA HA	Cn0,7% AA CHA HA	Cn1,0% AA CHA HA	Cn85% AA CHA HA
n-3	-0.37 -0.45 -0.06	-0.50 -0.46 -0.93**	-0.80 -0.18 -0.55*	-0.58* -0.42 0.15
n-6	-0.42 -0.32 0.01	-0.72** -0.77** -0.91**	-0.72** -0.18 -0.76**	-0.58* -0.50 -0.40
n-6/n-3	-0.01 0.32 0.11	-0.19 -0.62* -0.58*	0.01 -0.01 -0.13	0.27 0.04 -0.24
n-3/n-6	-0.38 -0.38 0.07	0.23 0.57* 0.58*	-0.05 -0.03 0.11	-0.22 0.11 0.34
C18:2/C18:3	-0.02 -0.60 -0.07	-0.45 -0.32 -0.60*	-0.48 -0.40 -0.25	0.11 -0.10 -0.07
C20:4/C20:5	-0.48 0.19 0.01	-0.45 -0.42 -0.14	-0.22 0.41 -0.34	0.18 -0.18 -0.37
SFA%	0.29 0.57* -0.03	0.33 0.22 0.02	0.47 0.38 0.77**	0.20 0.34 0.25
MUFA%	-0.03 -0.13 -0.02	0.14 0.64* 0.74**	0.30 -0.38 0.45	-0.28 -0.08 0.02
PUFA%	-0.49 -0.39 -0.01	-0.69* -0.67* -0.94**	-0.76** -0.22 -0.75**	-0.60* -0.52 -0.37
P/S	-0.46 -0.44 -0.01	-0.71** -0.70** -0.90**	-0.77** -0.38 -0.74**	-0.58* -0.49 -0.37
MUFA/SFA	-0.28 -0.40 -0.02	-0.10 0.53 0.62	-0.10 -0.42 -0.39	-0.34 -0.33 -0.15
C18:2/CLA	-0.02 -0.24 0.19	-0.34 -0.41 -0.57*	-0.30 -0.27 -0.43	0.20 -0.07 -0.07
C18:3/CLA	-0.04 -0.02 0.10	-0.38 -0.50* -0.74**	-0.44 0.57* -0.60*	-0.37 -0.19 -0.26
AI	0.20 0.74** 0.31	0.12 -0.16 0.14	0.05 0.28 0.82**	0.58* 0.44 0.21
TI	-0.04 -0.04 0.06	-0.65* -0.63* -0.89**	-0.62* 0.19 -0.69**	-0.14 -0.33 -0.13
Thiostearate	0.15 -0.29 -0.17	-0.04 -0.21 -0.58*	0.45 0.79** -0.67*	-0.70** -0.33 0.06

Note: * $P < 0.05$ **$P < 0.01$ SFA (C14:0 + C16:0 + C18:0); MUFA (C16:1 + C18:1); n-3 (18:3 + 20:5 + 22:5 +2 2:6); n-6 (18:2 + 18:3 + 20:3 + 20:4 + 22:4); PUFA% = n-6 + n-3%.

Table 7. Effects of diet and genotype on enzymatic indices in intramuscular fat content of *Longissimus dorsi* muscle. Δ 9 desaturase: (C16:1/C16:0 and C18:1/C18:0); elongase (C18:1/C16:1); Δ 5 desaturase (C20:4/C20:3) and combined effects of desaturases and elongases (C20:3/C18:2 and C20:4/C18:2) in n-6 PUFA and elongase, C22.5/C20:5) and combined effects of desaturases and elongases (C20:5/C18:3 and C22:5/C18:3) involved in n-3 PUFAs metabolism

Item	Genotype AA CHA HA			Diet Pasture Cn0.7% Cn1% Cn85%				RESM	Genotype	Diet	GxD
C16:1/C16:0	0.14	0.14	0.17	0.14	0.15	0.15	0.16	0.02	0.001	0.0029	0.022
C18:1/C18:0	2.59b	2.68b	3.15a	2.49c	2.75b	2.89ab	3.09a	0.31	0.001	0.001	NS
C18:1/C16:1	10.18a	10.58a	9.13b	9.96a	10.13a	10.06a	9.51b	1.12	0.001	0.001	NS
C20:4/C20:3	2.70b	2.87ab	3.31a	2.72b	2.98ab	2.97ab	3.18a	1.02	0.01	NS	NS
C20:3/C18:2	0.11	0.10	0.11	0.12	0.11	0.11	0.09	0.03	NS	0.001	0.05
C20:4/C18:2	0.27b	0.27b	0.34a	0.32a	0.31a	0.31a	0.24b	0.09	0.001	0.001	NS
C22:5/C20:5	1.82	2.04	1.96	1.38b	1.76b	2.04ab	2.58a	1.17	NS	0.001	NS
C20:5/C18:3	0.36ab	0.33b	0.41a	0.42a	0.34c	0.34c	0.38b	0.13	0.05	0.05	NS
C22:5/C18:3	0.62ab	0.59b	0.72a	0.53b	0.56b	0.61b	0.84a	0.24	0.05	0.001	NS

Note: Significance of main effects (genotype and diet) and their interactions (G*D) a, b, means within a row with different letters significantly differ (P < 0.05). REMS (Root mean square error).

The index of Δ^9 desaturase enzyme activity on the conversion of C16:0 to C16:1 and C18:0 to C18:1 increased as the grain in the diet increased and was higher in HA than in the other genotypes. The lower values of the desaturase and higher values for CLA in P (Table 3) could be explained according to Daniel et al., (2004) who stated that the increased concentrations of CLA in forage rich diets are associated with an increase in substrate (C18:1 trans 11) availability and not with an increase in SCD gene expression. The HA genotype presented the higher values for both Δ^9 desaturases which may explain the genetic basis for breed differences ($p < 0.05$) in C16:1 and C18:1 between HA and AA. This was mainly due to increases in C18:1/C18:0. Δ 9 desaturases C16:1/C16:0 and C18:1/C18:0 where affected by diet (Cn85% = Cn1% > Cn0.7% > P) and genotype (HA > AA = CHA) (Table 7). The expression of stearoyl-CoA desaturase (SCD) is associated with adipocyte hypertrophy in a number of species. Therefore, depressing SCD enzyme activity may decrease carcass adiposity. Myristate, palmitate and stearate are converted to their (n-9) corresponding monounsaturated fatty acids by Δ^9 desaturase and palmitate is converted to stearate through chain elongation by elongase (Dance et al., 2009). Tissues with less desaturase activity (loin and round) are then more susceptible to differences in rumen outflow of fatty acids caused by changes in ruminal biohydrogenation. The expression and activity of tissue Δ^9 desaturase or steroyl-Coa desaturase (SCD) has been shown to be inhibited by PUFAs. The depressive effect of PUFAs on SCD increases with increases in the carbon chain length and double bond number within the number of double bonds within the PUFA (Ntambi, 2004). Several studies demonstrated the existence of polymorphism in the SCDI and SREBP-1 genes in the population of Fleckvich cattle and that the SCDI polymorphism was associated with the proportion of several FA indices in both muscle and subcutaneous fat tissues (Mele et al., 2007). Therefore, depressing SCD enzyme activity may decrease carcass adiposity. Vasta et al., 2009 also found that concentrate-fed lambs had higher MUFA/SFA and C18:1/C18:0 ratios, which indicate changes in Δ 9 activity or reflect differences in the diet composition. All ratios were affected significantly by diet and breed. Feedlot LD had higher ratios than pasture LD which indicate changes in Δ9 desaturase activity or reflect differences in the diet composition. Variations in Δ9 activity might be related to the changes in expression of Δ9d mRNA and/or protein but the information about dietary regulation of Δ9d in ruminants is very limited (Vasta et al., 2009). The elongase C18:1/C16:1 index was affected by genotype (AA = CHA > HA) and diet (Cn85% < P = Cn07% = Cn1%). The elongase (C18:1/C16:1) was lower in genotype HA and in Cn85% diet. Dance et al., (2009) demonstrated that there are breed-specific and tissue-specific variations in CLA, MUFA, total fat content and SCD expression in beef cattle. The variations in SCD protein expression might contribute to the breed-specific variations in MUFA and CLA content. The indices of Δ 5 desaturase (C20:4/C20:3) were only affected by genotype (HA = CHA > AA); The indices of combined effects of desaturases and elongases were affected as it follows: C20:3/C18:2 was affected by diet (Cn85% > P = Cn0.7% = Cn1%); C20:4/C18:2 was affected by genotype (HA > AA = CHA) and diet (Cn85% < Cn1% = Cn0.7 = P); C22:5/C20:5 was affected by diet (Cn85%= Cn1% > Cn0.7% = P); C20:5/C18:3 was affected by genotype (HA = AA > CHA) and diet (P > Cn0.7% = Cn1% = Cn85%); C22:5/C18:3 was affected by diet (Cn85% > Cn1% = Cn0.7% = P) and genotype (HA = AA > CHA).

The correlation of enzymatic indices with IMF content are presented in Table 8. The correlations were very variable: C16:1/C16:0 (r = 0.07 to 0.80**), C18:1/C18.0 (r = 0.09 to 0.62*) and C18:1/C16:1 (r = 0.04 to 0.58*). Corazzin et al., (2013) showed no differences in the Δ^9 desaturase index when comparing animals with different subcutaneous fat depots, but in contrast with the finding of Jan et al., (2008), who reported a positive correlation between Δ^9 desaturase activity and the marbling in beef. The desaturase, C20:4/C20:3 (r = 0.02 to -0.86**); C20:3/C18:2 (r = 0.02 to -0.71), C20:4/18:2 (r = -0.09 to 0.73**), C22:5/C20:5 (r = 0.06 to 0.75*), C20:5/C18:3 (r = 0.9 to-0.71) and 22:5/C18:3 (r = -0.10 to -0.58*).

Table 8. Pearson correlation coefficients between some enzymatic indices and intramuscular fat content of *Longissimus dorsi* muscle. Δ 9 desaturases: (C16:1/C16:0 and C18:1/C18:0); elongase (C18:1/C16:1); Δ 5 desaturase (C20:4/C20:3) and combined effects of desaturases and elongases (C20:3/C18:2 and C20:4/C18:2) in n-6 PUFA and elongase, C22.5/C20:5) and combined effects of desaturases and elongases (C20:5/C18:3 and C22:5/C18:3) involved in n-3 PUFAs metabolism

Item	Pasture AA CHA HA	Cn0.7% AA CHA HA	Cn1.0% AA CHA HA	Cn85% AA CHA HA
C16:1/C16:0	-0.32 0.07 -0.06	0.29 0.35 0.39	0.07 -0.80** 0.29	-0.39 -0.12 -0.48
C18:1/C18:0	-0.09 -0.42 -0.16	0.27 0.62* 0.37	0.13 -0.57* 0.00	-0.25 -0.15 -0.18
C18:1/C16:1	0.06 -0.25 0.19	-0.36 -0.06 0.00	-0.32 0.58* -0.36	0.04 -0.11 0.48
C20:4/C20:3	-0.03 0.37 0.20	-0.58* 0.02 0.16	-0.86** -0.06 -0.58*	-0.74** -0.48 -0.30
C20:3/C18:2	-0.25 -0.32 -0.40	0.44 -0.48 0.01	-0.71** -0.36 0.12	0.68* -0.02 -0.37
C20:4/C18:2	-0.38 -0.30 -0.27	0.45 -0.66* 0.07	-0.73** -0.09 -0.46	-0.48 -0.68* -0.61*
C22:5/C20:5	-0.38 0.09 -0.13	0.10 0.59* 0.22	-0.63* 0.75** -0.23	0.51 0.11 -0.06
C20:5/C18:3	-0.36 -0.18 -0.26	-0.27 -0.46 -0.63	-0.13 -0.71** -0.08	-0.26 0.09 -0.16
C22:5/C18:3	-0.48 -0.10 -0.57*	-0.25 0.38 -0.15	-0.23 -0.46 -0.36	0.32 -0.13 -0.13

Note: * significant correlation ($P < 0.05$) ** significant correlation ($P < 0.01$).

3.4. Effects of Diet, Genotype and IMF Content on Cholesterol Content

The effects of genotype and diet on total cholesterol (mg/100g muscle) are presented in Table 5. Diet and genotype affected significantly the cholesterol levels P < 0.001. Many authors consider that muscle cholesterol concentrations do not vary in response to the differences in breed type, gender class or diet (Brugiapaglia et al., 2014). Maybe changes in cholesterol content in muscle may require marked changes in structure of muscle cells associated with a marked redistribution of membrane fatty acids.

In Table 9 are presented the correlation of total cholesterol and specific cholesterol (mg/g) with IMF%. No significant correlations were detected of total cholesterol with IMF%. On the contrary specific cholesterol was negatively correlated with IMF%. A clear dilution effect of cholesterol in the intramuscular fat seems to be evident. The specific cholesterol values were higher in pasture beef in the three genotypes compared with the supplemented and feedlot beef.

The specific cholesterol content seems to be more sensible to animal dietary changes than the total beef cholesterol. The relation between beef intramuscular fat and specific cholesterol could contribute to understand the erratic results related to cholesterol content in beef under different production systems.

Public concern over the effects of dietary cholesterol on heart disease is more specifically related to meat products, especially red meat. Several researches reported the lower cholesterol in bulls and steers finishing in pasture systems as opposed to traditional finishing (Morales et al., 2012). Others considered that breed; nutrition and gender do not affect cholesterol concentrations of bovine skeletal muscle. The results from these previously mentioned studies indicated an unpredictable relationship between fatness and cholesterol. Cattle category was found to influence significantly (P < 0.07) cholesterol level. Litwincz, et al., (2015) found the highest cholesterol concentrations in cows. In other categories cholesterol level was significantly lower ranging from 44.30 to 57.93 mg/100g. In line with results found by Bures et al., (2006), Costa et al., (2006) and Garcia et al., (2008). However, Desimone et al., (2013) found cholesterol ranging from 62.55 to 76.60 mg/100g in commercial beef cuts according to USDA quality grade.

Table 9. Effects of genotype and diet in total (mg/100g LD) and specific (mg/g IMF) cholesterol and Pearson correlations with IMF content

	IMF%	Cholesterol mg/100g	Correlation mg/100g with IMF	Cholesterol mg/g IMF	Correlation mg/g IMF with IMF
Pasture					
AA	3.05b	44.1a	0.13a	15.4a	-0.88**
CHA	2.71b	43.0a	0.08a	15.7a	-0.68*
HA	2.70a	44.4a	0.15a	17.2a	-0.72**
Cn0.7%					
AA	3.6a	44.2a	0.15a	13.8b	-0.86**
CHA	4.0a	47.3a	0.05a	12.7b	-0.78**
HA	3.4a	44.8a	0.14a	14.0b	-0.81**
Cn1%					
AA	3.4a	46.3a	0.12a	14.1b	-0.92**
CHA	4.3a	45.3a	0.11a	11.6b	-0.57*
HA	4.4a	42.5a	0.06a	10.0b	-0.79**
Cn85%					
AA	4.1a	44.1a	0.39a	11.3b	-0.83**
CHA	3.9a	46.4a	0.06a	14.2b	-0.93**
HA	3.8a	46,7a	0.12a	12.9b	-0.87**

* Significant correlation (P < 0.05); ** significant correlation (P < 0.01).

CONCLUSION

The fatty acid profile shows a wide variability depending on factors such as genotype and diet. The overall intramuscular fat content has a variable effect on relative fatty acid composition and, consequently, with nutritional and enzymatic indices. The intramuscular fat content impacts on the relative fatty acid composition of neutral and phospholipids, with implications for the nutritional quality of beef.

The actual produced beef showed higher variability for fat content and fatty acid profile, and the general message that the beef is unhealthy, due to being rich in IMF and SFAs is misleading for the consumer. Therefore, more detailed information on meat composition may be useful to help consumer decision making.

It is well known that fatty acid composition of beef, and hence its nutritional value, can be manipulated by genetic and nutritional approaches, although it is acknowledged that genetic factors provide smaller differences than dietary factors. According to Wood et al., (2008), the results obtained in this study confirm that variation in IMF content has a fundamental effect on FA composition. While the concentration in phospholipids, rich in PUFA, is nearly constant and relatively independent from the total IMF amount, the IMF, consistent mainly of triacylglycerols, is associated with a higher IMF level.

REFERENCES

Alfaia, C. P. M., Alves, S. P., Martins, S. I. V., Costa, A. S. H., Fontes, C. M. G. A., Lemos, J. P. C., Bessa, R. B., Prates, J. A. M. (2009). Effect of the feeding system on intramuscular fatty acids and conjugated linoleic acid isomers of beef cattle, with emphasis on their nutritional value and discriminatory ability. *Food Chemistry*, 114, 939-946.

AOAC 1992. *Official Methods of Analysis* 15th edition. 3er Suppl. International Gaithersburg, MD. pp 19-140

Bellizi, M. C., Franklin, M. F., Duthie, G. G., James, W. P. T. (1994). Vitamin E. and coronary heart disease: The European paradox, *European Journal Clinical Nutrition* 48; 822-831.

Bauchard, D., Gladine, C., Gruffat, D., Leloutre, L., Durand, D. (2005). Effects of diet supplemented with oil seeds and vitamin E on specific fatty acids of Rectus abdominis muscle in Charolais fattening bulls. In

Hocquette &S. Gigli (Eds). *Indicators of milk and beef quality*, 112,431-436 EAAP Publ., Wageningen Academic Publishers.

Brugiapaglia, A., Lussiana, C., Destefanis, G. (2014). Fatty acid profile and cholesterol content of beef at retail of Piamontese, Limousin and Frisian breeds. *Meat Science 96*, 568-573.

Bures, D., Bartoñ, L., Teslik, V., Zahradkova, R., (2006). Chemical composition, sensory characteristics, and fatty acid profile of muscle from Aberdeen Angus, Charolais, Simmental and Hereford bulls. *Czech Journal of Animal Science*, 51, 279-284.

Choi, Y. J., Prak, Y. H., Michael, W. P., Ntambi, J. M. (2001). Regulation of stearoyl CoA desaturase activity by the trans 10 cis-12 isomer of conjugated linoleic acid in HepG2 cells. *Biochem. Biophys Res. Commun*. 284, 689-693. Doi:10.1006/bbrc.2001.5036.

Corazzin, M., Bovolenta, S., Sepulcri, A., Piasentier, E. (2012). Effect of whole linseed addition on meat production and quality of Italian Simmental and Holstein young bulls. *Meat Science*, 90, 99-105.

Costa, P., Roseiro, L. C., Partidario, A., Alves, V., Bessa, R. J. B., Calkins, C. R., Santos, C. (2006). Influence of slaughter season and sex on fatty acid composition, cholesterol and alpha-tocopherol contents on different muscles of Barrosa-PDO veal. *Meat Science*, 72, 130_139.

Daley, C. A., Abbott, A., Doyle, P. S., Nader, G. A., Larson, S. A. (2010). Review of fatty acid profiles and antioxidant content in grass-fed and grain-fed beef. *Nutrition Journal,* 9, 1-12.

Dance, I. J. E., Matthews, K. R., Doran, G. (2009). Effect of breed on fatty acid composition and stearoyl-CoA desaturase protein expression in the *Semimembranosus* muscle and subcutaneous adipose tissue of cattle. *Livestock Science*, 125, 291-297.

Daniel, Z. C., Wynn, R. J., Salter, A. M., Buttery, P. J. (2004). Differing effects of forage and concentrate diets on the oleic acid and conjugated linoleic acid content of sheep tissues: the role of stearoyl-CoA desaturase. *Journal Animal Science*, 82,747-758.

Desimone, T. L., Acheson, R. A., Woerner, D. R., Engle, T. E., Douglass, L. W., Belk, K. E., (2013). Nutrient analysis of the Beef Alternative Merchandising cuts. *Meat Science*, 93,733-745.

148 *Pilar T. Garcia, Nestor N. Latimori, Ana M. Sancho et al.*

FAO. Food and Agricultural Organization for the United Nations. *Fats and fatty acids in Human Nutrition. A report of an expert consultation.* Roma: FAO 2010.

Folch, J., Lees, M., Sloane-Stanley, G. H. S. (1957). A simple method for the isolation and purification of total lipids from animal tissues. *Journal Biological Chemistry,* 226, 497-509.

Food and Agriculture Organization for the United Nations. (2010). Fats and fatty acids in Human Nutrition: *Report of an expert consultation, FAO Food and Nutrition* Paper 91, FAO. Rome.

Garcia, P. T., Pensel, N. A., Sancho, A. M., Latimori, N. J., Kloster, A. M., Amigone, M. A., Casal, J. J. (2008). Beef lipids in relation to animal breed and nutrition in Argentina. *Meat Science*, 79, 500-508.

Garcia, P. T., Casal, J. J. (2012). Linoleic and alfa-linolenic acids content of pork, beef and lamb lipids. In *Linoleic acid: Sources, Biochemical Properties and Health Effects.* Editor Igho Onakpoya, Nova Science Publishers 1st edition ISBN 978-1-62257-399-8.

Garcia, P. T., Pordomingo, A., Perez, C. D., Rios, D. M., Sancho, A. M., Volpe, Casal, J. J. (2015). Influence of cultivar and cutting date on the fatty acids composition of forage crops for grazing beef production in Argentine. *Journal Grass Forage Science.* DOI:10:111/GFS 12167.

Garcia, P. T., Latimori, N., Sancho, A. M., Casal, J. J. (2017). Diet and Genotype Effects on n-3 Polyunsaturated Fatty Acids of Beef Lipids. *Research in Agriculture,* ISSN 2740-4431. 2, N°1.

Indurain, G., Berlain M. J., Goñi, M. V., Arana, A., Purroy, A. (2006). Composition and estimation of intramuscular and subcutaneous fatty acid composition in Spanish young bulls. *Meat Science*, 73,326-334.

Jiang, Z., Michal, J. J., Tobey, D. J., Daniels, T. F., Rule, D. C., MacNeil, M. D. (2008). Significant associations of stearoyl-CoA desaturase (SCD1) gene with fat deposition and composition in skeletal muscle. *Internal Journal Biological Science*, 4, 345-351.

Kramer, J. K. G., Cruz-Hernandez, C., Deng, Z., Zhou, J., Jabreis, G., Dugan, M. E. R. (2004). Analysis of conjugated linoleic acid and thans 18:1 isomers in synthetic and animal products. *American Journal Clinical Nutrition*, 79,1137S-1145S.

The Relationship of Fatty Acid Composition ...   149

Laborde, F. L., Mandell, I. B., Tosh, J. J., Wilton, J. W., Buchanan-Smith, J. G. (2001). Breed effects growth performance, carcass characteristics, fatty acid composition, and palatability attributes in finishing steers. *Journal Animal Science,* 79,355-365.

Lee, M. R. F., Evans, P. R., Nute, G. R., Richardson, R. I., Scollan, N. D. (2009). A comparison between red clover silage and grass silage feeding on fatty acid composition, meat stability and sensory quality of the M. *Longissimus muscle* of dairy cull cows. *Meat Science,* 81, 738-744.

Litwincezuk, Z., Domaradzki, P., Grodzicki, T., Litwincezuk, A., Florek, M., (2015). The relationship of fatty acid composition and cholesterol content and marbling in the meat of Polish Holstein-Friesian cattle from semi-intensive farming. *Animal Science Papers and Reports,* 33. 119-127.

Malau-Aduli, A. E. O., Siebert, B. D., Bottema, C. D. K., Pitchford, W. S. (1998). Breed comparison of the fatty acids composition of muscle phospholipids in Jersey and Limousine cattle. *Journal of Animal Science,* 76, 766-773.

McAfee, A, McSorley, E. M., Cuskelly, G. J., Moss, B. W., Wallace, J. M. W., Bonham, M. P., Fearon, A. M. (2010). Red meat consumption: An overview of the risk benefit. *Meat Science,* 84,1-13.

Mele, M., Conte, G., Castiglioni, B., Chessa, S., Macciotta, N. P. P., Serra, A., Buccioni, A., Pagnacco, G, Secchiari, P. (2007). Stearoyl-CoA desaturase gene polymorphism and milk fatty acid composition in Italian Holsteins. *Journal of Dairy Science,* 90.4458-4465.

Mensing, R. P. (2005). Effects of stearic acid on plama lipid and lipoproteins in human. *Lipids,* 40,1201-1205.

Morales, R., Folch, C., Iraira, S., Teuber, N., Realini, C. E. (2012). Nutritional quality of beef produced in Chile from different production systems. *Chilean Journal of Agricultural Research,* 72, 80-86.

Nogalski, Z., Wielgosz-Groth, Z., Purwin, C., Sobczuk-SZul, M., Mochol, M., Pogorzelska-Przybylek, P., Winarski, R. (2014). Effect of slaughter weight on the carcass value of young crossbred (Polish

150 *Pilar T. Garcia, Nestor N. Latimori, Ana M. Sancho et al.*

Holstein Friesian x Limousin) steers and bulls. *Chilean Journal of Agricultural Research*, 74, 59-66.

Ntambi, J. M., Miyazaki, M., (2004). Regulation of stearoyl- CoA desaturases and role in metabolism. *Progress in Lipid Research*, 43, 91-104.

Pariza, M. W., Park, Y., Cook, M. E. (2001). The biologically active isomers of conjugated linoleic acid. *Progress of Lipid Research,* pp 283:298.

Pawloski, R. J., Hibbeln, J. R., Novotay, J. A., Salem, N. Jr (2001). Physiological compartamental analysis of alfa-linolenic acid metabolism in adult humans. *Journal Lipid Research,* 42, 1457-1265.

Raes, K., De Smet, S, Demeyer, D. (2001). Effect of double-musculing in Belgian Blue young bulls on the intramuscular composition with emphasis on conjugated linoleic acid and polyunsaturated fatty acids. *Animal Science,* 73,253-260.

Ulbrich, T. L. V., Southgate, D. A. T. (1991). Coronary heart disease: seven dietary factors. *Lancet* 338, 985-992.

Vasta, V., Priolo, A., Scerra, M., Hallet, K. G.,Wood, J. D.,Doran, O. (2009).Δ9 desaturase protein expression and fatty acid composition of longissimus dorsi muscle in lamb fed green herbage or concentrate with or without added tannins. *Meat Science*, 82,357-364.

Wood, J. C., Enser, M., Fisher, A. V., Nute, G. R., Sherd, R. I., Richardson, R. I., Hughes, S. I., Whittington, F. M. (2008). Fat deposition, fatty acid composition and meat quality: A review. *Meat Science,* 78,343-358.

Yu, S., Derr, J., Etherton, T. D., Kris-Etherton, P. M. (1995). Plasma cholesterol–predictive equations demonstrate that stearic acid is neutral and monounsaturated fatty acids are hypocholesterolemic. *American Journal Clinical Nutrition*, 61, 1129-1139.

Zhang, S., Knigh, T. J., Stalder, K. J., Goodwin, R. N, Lonergan, S. M., Beitz, D. C. (2007). Effects of breed, sex, and halothane genotype on fatty acid. *Journal Animal Sci*ence, 85,583-591.

In: Beef: Production and Management Practices ISBN: 978-1-53613-254-0
Editor: Nelson Roberto Furquim   © 2018 Nova Science Publishers, Inc.

Chapter 6

THE BRAZILIAN GREEN BEEF: THE IMPORTANCE OF PASTURE MANAGEMENT FOR ANIMAL PERFORMANCE AND QUALITY OF THE MEAT

Bruno Lala[1,], Vinícius Valim Pereira[2], PhD,*
Ulysses Cecato[3], PhD,
Guilherme Sicca Lopes Sampaio[1], PhD,
Ana Paula Possamai[4], PhD and Ana Maria Bridi[5], PhD

[1]Department of Economy, Sociology and Technology,
São Paulo State University, Botucatu, SP, Brazil
[2]Department of Engneering,
Faculdades Pitágoras, Divinópolis, MG, Brazil
[3]Department of Animal Science
State University of Maringá, Maringá, PR, Brazil
[4]Department of Animal Science,
Faculdades Unidas Vale do Araguaia, Barra do Garças, MT, Brazil
[5]Department of Animal Science
State University of Londrina, Londrina, PR, Brazil

---

[*] Corresponding Author Email: brunolala@hotmail.com.

152 Bruno Lala, Vinícius Valim Pereira, Ulysses Cecato et al.

ABSTRACT

Brazil possesses circa 90 million hectares of good quality pastures and 15 million hectares of pastures that are currently in state of degradation. However, according to a report on changes in land cover and use in Brazil, published in 2014, the total area of cultivated pastures saw a reduction in its growth rate from 11.1% in 2010/2012 to 4.5% in the following period. The amount of degraded pasture land in Brazil is estimated to be at around 100 million hectares, including both planted and natural pastures. With the increase in herds and reduction of the planted area, it is necessary to use and improve pasture management techniques in order to increase meat production in Brazil. In fact, current productivity of Brazilian beef cattle, estimated at 5.1 @/ha/year, is considered low. It is a well-known fact that that there is a possibility to improve the efficiency of domestic livestock farming. An increase of just around 20% in the productivity of planted pastures in Brazil would be enough to satisfy the demand for meat, grains, timber products and biofuels for the next 30 years, without the need to incorporate new areas and thus destroy natural ecosystems. This could be done through rational intensification of management process for existing pasture areas, with an emphasis on the recovery of degraded pastures. It is possible to liberate areas for other production activities, reducing deforestation without compromising the supply of food to the population. Therefore, as the territorial, climatic and socioeconomic characteristics of the country predetermine the fact that in the next decades, pastures will continue to play a fundamental role in Brazilian cattle raising, the correct management of pastures and better adaptation to current grazing systems in addition to the genetic quality of beef cattle herds, tend to secure the role of Brazil as a reference in the production of pasture cattle-derived meat. Beef is a healthy source of vitamins, minerals, protein and energy, and the world market on average sells 3 million tons of fresh, chilled and processed meat, which accounts for 26% of world production. Recently, Brazil has become the largest exporter of beef, and the local production is based on animals raised on pasture. *Bos taurus indicus* is the most widely used genetic group in the country, due to their excellent adaptation to local climatic conditions and fodder. Data from the USDA's Annual Livestock Report shows that by 2017, Brazil's livestock population will reach 226,037 million heads, and of these, 85% will be pasture cattle. This allows for lower production costs, apart from the supply of a healthy product with high nutritional quality and great marketing appeal - the so-called "green beef" or, alternatively, the "grass beef." The pasture-based production, whose main concern is production of quality meat, depends on the quantity and nutritional value of the fodder offered to the animals. Thus, pasture management leads to an increase in meat quality and higher animal performance, while being more beneficial to human health due to an increase in the amount of good fatty acids.

Keywords: animal production, grass beef, meat quality, pasture management, sustainability

1. INTRODUCTION

Brazil has the world's largest commercial cattle herd of 208 million heads, with 29.67 million animals slaughtered in 2016 (IBGE, 2017). According to the Annual Livestock USDA Report (USDA, 2016) it is expected that in 2017, Brazilian herd reaches 226.037 million animals, with 85% of cattle fed on pasture. The pasture-based beef production is a competitive and efficient way to produce good quality meat at low cost (Da Silva, 2009) and there is a continuing trend of increasing the area of cultivated pastures (African grasses), specially *Brachiaria* spp. and *Panicum* spp. (Ferraz and Felício, 2010), which are more productive, well adapted to the Brazilian soil, and are of a higher quality than the native Brazilian species.

Meat quality and carcass conformation are linked to fat deposition that occurs during the growth of the animal, as well as fat thickness and body weight before slaughter, the latter being a parameter used for the classification and payment for carcasses in Brazilian slaughterhouses (Fugita et al., 2012; Missio et al., 2013). A carcass with good quality should provide sufficient amount of fat to ensure its preservation and desirable characteristics for consumption (French et al., 2000), which vary according to the diet of animals. Intramuscular fat (marbling) is considered the main determinant of meat quality (Strydom et al., 2000), besides imparting juiciness and flavor to beef.

Beef is a healthy source of vitamins, minerals, protein and energy. An average of 13 million tons of fresh, chilled and processed meat is sold on the world market each year, which accounts for 26% of world production (ANUALPEC, 2013). In this chapter, we will discuss two species of forage

154 *Bruno Lala, Vinícius Valim Pereira, Ulysses Cecato et al.*

used in Brazil for pasture-based animal production, reflected directly in several aspects of carcass quality and bovine meat, and highlight the characteristics of fatty acids, as well as the importance of conjugated linoleic acid (CLA) for the creation of pasture cattle.

2. MAIN FORAGE SPECIES USED IN BRAZIL

In Brazil, pasture cattle systems have long and even permanent cycles (continuous grazing). This fact, associated with the seasonal variation inherent to the climate itself and the phenological stages of forage plants, admits characteristics related to the pasture structure, described as the spatial arrangement of aerial biomass (Gonçalves et al., 2009), which has relevant aspects regarding the development of cattle fed with grass.

2.1. *Brachiaria Brizantha* cv. Xaraés (Xaraés grass)

The *Brachiaria* species were introduced to Brazil in 1952 and currently occupy the largest area of pastures used for cattle breeding (Fonseca et al., 2010). Pastures have always been the main food base for cattle production in Brazil and there is a continuing trend to increase cultivated pasture areas, which have become more productive and of superior quality. Of the 170 million hectares of pasture, 100 million are cultivated and 70 million grow naturally; 80% of this cultivated area is formed by the *Brachiaria* genus, mainly in the Central-West Region (Valle et al., 2001; Fonseca et al., 2010). The states of Goiás and Mato Grosso do Sul, which have approximately 73 million hectares of pasture, have 51% of their respective areas formed by *Brachiaria*, 42% by native species and only 8% by other species (Macedo, 1995).

The use of *Brachiaria brizantha* cv. Xaraés was launched by EMBRAPA GADO DE CORTE Company in 2003 as a fitting option for the Marandu cultivar, especially for its greater support capacity and tolerance to soil flooding (Andrade and Assis, 2008), and as a new option for the diversification of forage grasses. Despite providing animal production inferior to that of other species, Marandu cultivar has some advantages such as greater regrowth speed and greater production of the forage mass, which guarantees higher support capacity and higher yield per area (Euclides et al., 2005). In order to improve the performance of the crop, it is necessary to increase the yield of the forage mass. The *Brachiaria brizantha* cv. Xaraés, originally from Burundi, Africa, was introduced in Brazil in the 1980s through a cooperation agreement signed between EMBRAPA and the International Center for Tropical Agriculture (CIAT). After an evaluation period that lasted for 15 years (according to the stages of the *Brachiaria* Genetic Improvement Program coordinated by EMBRAPA GADO DE CORTE), it received the code BRA 004308 and is now registered in the National Register of Cultivars of the Ministry of Agriculture, Livestock and Supply under the number 04509.

Few scientific works report these aspects, making it necessary to generate information about animal production using the Xaraés cultivar in order to obtain proper management guidance. Galbeiro (2009), in his study using different grades of grazing intensity with the Xaraés grass, concluded that the forage production and structural characteristics of the pastures point at a high degree of flexibility in handling offered by the Xaraés grass, with heights between 30 and 45 cm being the best, depending on forage production goals and animal performance targets.

In the case of Xaraés grass, another interesting factor worth mentioning is the forage accumulation pattern that results from the grazing strategy used. Da Silva and Nascimento Júnior (2007) describe the best height for handling this grass as being 30 cm, which produces 3000 kg ha-1 of forage mass on average. The grazing frequency, if adequately defined, ensures large forage production and good nutritive value, resulting in an increase in animal performance. Carloto et al. (2011), having investigated different heights of Xaraés grass, affirmed that the levels of crude protein and IVDM (*in vitro* Dry Matter Digestibility) were higher for pasture managed at 15 cm compared to that of 45 cm, while pasture at 30 cm showed values similar to those for other heights, which led to the conclusion that the lowest nutrient value of the tallest pasture was probably a consequence of the higher number of old leaves present in the canopy, since leaves rejected by the animals continue to age.

For animal production to be efficient and competitive in pasture systems, the forage plant must be used in a rational way through sustainable management practices that allow high productivity and efficient use of the forage produced, thus generating maximum animal productivity (Gomide and Gomide, 2001). The pasture provided to the animals is the determining factor in the polyunsaturated fatty acid content of the meat. The choice of forage species, the physiological stage, the availability of nutrients and pasture management can significantly alter this content; leaf ratio usually decreases with the aging of a plant and the stem has only about 1/2 - 1/3 of fatty acid content usually found in leaves (Jarrige et al., 1995).

Nitrogen fertilization may also increase omega-3 concentration in grasses, since forage lipids are found predominantly in the leaves, and the increased availability of nitrogen to the plant induces leaf biosynthesis (Harfoot, 1981). French et al. (2000) showed that the inclusion of fresh forage in the diets can provide greater deposition of Conjugated linoleic acid (CLA) in the tissues. This was not observed when the authors used maize silage as fodder.

2.2. *Panicum Maximum* cv. Tanzania

The *Panicum maximum* species originated in Africa and was probably introduced to the Americas in the late eighteenth century by sheer accident, when it was used as bed in slave ships that brought slaves to Brazil from the western part of the African continent (Jank, 1995). Subsequently, a dispersion of seeds by wind, birds and humans occurred in several regions of the country (Aronovich, 1995).

The Tanzania grass (*Panicum maximum* Jacq., Tanzania-1) was collected by the Office de la Recherche Scientifique et Technique d'Outre-Mer (ORSTOM) in Korogwe, Tanzania. It was then evaluated in several conditions after a long selection work coordinated by EMBRAPA GADO

DE CORTE in Campo Grande/MS and commercialized in 1990 (Jank, 1995). The grass is a cespitose plant with average height of 1.3 m in free growth and curved leaves with an average width of 2.6 cm. The stems are purplish, non-cereous, with glabrous sheaths and laminas. The inflorescences are panicle-shaped, with long primary branches and secondary branches that are long only at the base. The spikelets are purplish, glabrous and uniformly distributed. The whorls are glabrous (Savidan et al., 1990).

This cultivar has shown great acceptance by technicians and producers due to its high production potential (Penati, 2002) and forage quality, which are generally superior to those of *Brachiaria* (Euclides et al., 1995a). These characteristics provide high stocking rates and weight gains in pasture animals when soil fertility and management are appropriate for the requirements of the cultivar.

Tropical forage plants, such as Tanzania grass, should be evaluated thoroughly in terms of their nutritional qualities, with analysis of their bromatological composition requiring special attention and detail. Due to annual variations and differences during the development stage of Tanzania grass, a study of plant quality throughout the seasons is also necessary. Therefore, the use of fertilization is a strategy to minimize these variations in quality and to increase yield and productivity (Balsalobre, 2002).

The determination of the bromatological composition and the digestibility of the fractions that make up a forage plant are necessary in order to determine animal performance in ruminant production systems. Therefore, tropical forages, due to the seasonality of production, do not provide homogeneous amounts of nutrients for animal production throughout the year, with dry season providing the lowest yield and lowest nutritional quality (Euclides and Medeiros, 2003).

The main objective in the production systems is to maximize the dry matter intake of grazing animals, which is a key point in defining the balance of diets aiming at higher animal performance, since such definition of the consumption by grazing animals allows formulating rations, predict performance, estimate demand for food or animal requirements (Mertens, 1994).

It should be noticed that in grazing, the effect of selectivity is also of fundamental importance, as it directly influences the quality of the animal's diet. In addition, the nutrient composition of the forage is much more variable than that of any other food because it is influenced by the species and cultivar, by the chemical and physical properties of the soil, edaphoclimatic conditions, physiological stage and the applied handling techniques. Thus, the productivity of a pasture is determined by the concomitant effects of the above factors, and the response to these factors will vary for each species (Euclides and Medeiros, 2003).

The energetic value of forage can be determined by the digestibility of the organic matter and depends mainly on the degree of lignification of the cell wall (Paulino et al., 2006). When a plant reaches maturity, some changes in chemical composition occur (Balsalobre et al., 2001). The digestibility of fodder in ruminants is related to the distribution of lignin in the cells, the ratio between carbon and nitrogen, as well as the microbial population of the rumen. As the plant ages, the digestible components such as proteins, soluble carbohydrates and phosphorus tend to decrease, as lignin, cellulose, hemicellulose, cutin and silica content increases, causing the reduction of digestibility (Euclides et al., 1995b; Paulino et al., 2006). The stage of forage development directly influences the chemical composition and digestibility of the grass (Reis, 1993).

Tropical plants absorb CO_2 from the environment via the C4 cycle. These plants are characterized by high growth rates. With the advancement of physiological age, they lose their quality faster when compared to plants that predominate in temperate climates countries and use a different CO_2 absorption system called C3 (Van Soest, 1994). Nitrogen fertilization affects the nutritive value of forages, promoting variations in the chemical composition of the dry matter of plants (França et al., 2007). For Cecato et al. (2001), nitrogen provides an increase in crude protein ratio, a reduction in of neutral detergent fiber (NDF) levels and acid detergent fiber (ADF) levels in the dry matter of the resulting forage. Thus, the digestibility of the fiber is an important factor that influences the consumption of dry matter.

The Brazilian Green Beef 159

Indigestible fiber can occupy digestive tract by filling the rumen up to its capacity and, consequently, significantly reducing ruminal space and dry matter intake (Thiago and Gill, 1990).

3. ASPECTS OF BOVINE CARCASSES

The slaughter of cattle in Brazil reached a new historical record in 2014, with 42.255 million of animals slaughtered. This figure showed an increase of 5.3% in relation to the preceding quarter and 11.7% in comparison with the second quarter of 2012 (IBGE, 2014). According to Carvalho Junior et al. (2009), carcass evaluation is an important method for analysis of performance achieved by an animal during its development and is determined by examining consumption, weight gain, feed conversion and carcass yield. The meat production system is evaluated using quantitative and qualitative characteristics of the carcass; the quantitative characteristics are determined by yield, regional composition, tissue composition and muscle mass (Lucas, 2007). The quality of the meat and the conformation of carcasses are MORE related to the muscular development that occurs during the growth of the animal, as well as to the fat covering and body weight before slaughter, which is the parameter used for the classification of and payment for carcasses in Brazilian slaughterhouses. (Fugita et al., 2012; Missio et al., 2013).

Lush (1926) was the first researcher to conclude that the method of estimating carcass composition by obtaining cuts from all the ribs was quite accurate. After his pioneering work, Hopper (1944) proposed the use of cuts from the 9th to 11th rib section instead of the whole cut for a more precise evaluation and lower depreciation of carcasses. Later, the methodology of Hankins and Howe (1946) was established in order to determine the physical composition of the 9th to 11th rib cut using a technique that accurately estimates carcass composition of bovines (Jorge and Fontes, 2001).

According to Costa et al. (2002), carcass weight and yield are the variables most widely used for frozen product commercialization. While comparing males and females, Di Marco (1998), found better carcass weights in males and explained that the fact could be related to greater growth momentum caused by the androgenic hormones, mainly testosterone. Restle et al. (2002) noticed that warm carcass yield is the most important characteristic for the producer, since it is directly linked to the commercial value of the animal. For Brondani (2002), carcass yield is of paramount importance to refrigerator-based storage because it is related to muscular mass.

A good quality carcass must have enough fat to guarantee its preservation and desirable characteristics for consumption. Intramuscular fat (marbling) is considered the main determining factor of meat quality, as it imparts flavor, succulence and aroma to beef, and is an important component in carcass classification system and pricing for the North American producer.

4. ASPECTS OF BOVINE MEAT

The increasing consumer demand for food quality forces organizations to develop tools that are more efficient in terms of safety and differentiation, while adding more value to their products (Spers, 2003). Cattle production in pastures is a competitive and efficient way to produce good quality meat at low cost (Da Silva, 2009). Thus, the companies always seek to develop their own standards in order to guarantee better quality of their differentiated product and a healthier option with higher benefit to the final consumer (Barcellos, 2007). According to Saab (1999), the differentiation of beef generates consumer confidence in this product due to its superiority and guarantee of origin, meaning that the consumer-based approach prevails in order to maintain credibility in the marketed product.

In addition to product quality, in recent years we have seen consumers seeking more detailed information on the origin of products (Zawadzki et al., 2013). Softness, along with meat color, are important factors that affect the acceptability of the product by consumers (Igarasi, 2008). Grazing is

The technique used in most of the national territory, due to the availability of forage areas, low cost and the efficiency associated with this system.

The performance of pasture animals, expressed in production per animal unit, is dependent on various factors such as: genetics, forage intake, forage nutritive value and efficiency of consumed forage conversion (Gomide and Gomide, 2001). Pasture animals show darker-colored meat than those confined to feeding lots, and this difference is explained by different ages and amount of physical exercise, which increases the amount of myoglobin in the muscle (Bridi et al., 2011).

4.1. Fatty Acids

The consumption of animal fat is often linked to health problems, especially heart diseases. However, recent studies have shown that some isomers of fatty acids may actually have beneficial effects, thus evidencing the need for research that would evaluate the effects of fatty acids in an isolated manner. Consequently, there has been an increasing interest in possible ways of manipulating the fatty acid content in the meat by modifying the animal's diet.

Meat from grazing animals usually has a high concentration of polyunsaturated fatty acids (omega 3, C18:3), which is higher than recommended minimum values (Todaro et al., 2004). In general, silage contains higher levels of linolenic acid (C18:3), precursor of the omega 3 fatty acid series, while concentrate is richer in linoleic acid (omega 6, C18:2), precursor of the omega 6 series (Ponnampalam et al., 2001). Some studies have shown that cattle raised and fed on pasture present higher amounts of omega 3 in their meat, while those fed with grains have a higher proportion of omega 6 (Enser et al., 1998; French et al., 2000, Garcia et al., 2008; Bressan et al., 2011). That is due to the fact that grasses have a higher concentration of linolenic acids (omega 3), while the grains are rich in linoleic acid (omega 6).

According to World Health Organization (WHO, 2008), higher proportions of omega-3 in human diet is important to avoid the onset of coronary disease, autoimmune diseases, breast, prostate and colon cancer, as well as rheumatoid arthritis. However, few available studies offer characterization of pasture-derived beef. The results of the study on fatty acid profile of strictly pasture-fed cattle and those fed with different amounts of concentrate by French et al. (2000) showed that the meat of pasture-fed animals had lower percentages of saturated fatty acids, higher percentages of unsaturated fatty acids, better (lower) omega-6:omega-3 ratio and higher percentages of conjugated linoleic acid (CLA).

The fatty acid profile has significant importance in determining the physical, chemical and organoleptic properties of foods. As reported by Van Soest (1994), the composition of lipids varies from 2 to 4% in the dry matter of forage and consists of glycolipids and phospholipids, while the main fatty acids are linoleic and linolenic acids. At the same time, lipid content in seed oils is mainly triglycerides containing linoleic acid and oleic acid. The change in the fatty acid profile is interesting in terms of reducing the risks of coronary diseases, since the medium chain acids (lauric, myristic and palmitic acids) are hypercholesterolemic (Williams, 2000). Products of animal origin, such as beef, contain compounds beneficial to health, among which the CLA stands out, which presents anticarcinogenic properties and acts as nutrient delivery agent (McGuire and McGuire, 2000).

4.2. Conjugated Linoleic Acid (CLA)

Linoleic acid is the long chain fatty acid, which is considered essential, represented by C18: 2 (9,12). Conjugated linoleic acid (CLA) is not a single molecule, but a series of positional and geometric isomers of linoleic acid with double bonds that may be in the cis or trans form, separated only by a simple carbon bond (Medeiros et al., 2010). There are numerous isomers of linoleic acid, but C18: 2 cis-9, trans-11 (c9, t11), also known as rumanic acid, has anticancer properties and C18: 2 cis-12, trans- 10 (c12, t10), is a potent nutrient delivery agent, which means that the compound is capable of redirecting the destination of consumed nutrients, usually between fat synthesis and protein synthesis.

In dairy products, more than 80% of CLA is in the form of cis-9 and trans-11 isomers, which are biologically active forms of this acid. Foods from cattle, sheep and goats (meat and dairy products) usually present levels of CLA between 3 and 7 mg g-1 of fat, which may be raised by altering the diet of the animals (Ip and Pariza, 2005). Pork, fish, poultry and vegetable oils have lower amounts of CLA compared to ruminant derived products.

The effect of trans-10 cis-12 CLA is mainly related to changes in the lipid metabolism (Pariza, Park and Cook, 2001), and the ability to inhibit fat synthesis in the body, notably via the trans-10 cis-12 isomer, which has an effect on body composition. Some mechanisms have been proposed to explain this effect, among which the following can be mentioned: reduction of the esterification of fatty acids in triglycerides, interference of the cis-trans-10 isomer in differentiation of hepatocytes, reduction of lipogenesis and increase in lipolysis (Hayashi, 2003).

Using pastures generally leads to an increase of CLA content in the meat, when compared to the supply of complete feed mix or preserved fodder. Grass fed animals also have meat with higher amounts of CLA when compared to confined animals. This fatty acid has importance in nutrition and human health, since it can act as anticarcinogenic, antioxidant, antidiabetic and immunostimulatory agent (Bauman and Griinari, 2001). In a study carried out by Realini et al. (2004), intramuscular fat from grazing animals showed a higher concentration of total CLA content, as well as cis-9, trans-11 CLA isomers when compared to animals fed with concentrate (5.3 vs. 2.5 and 4.1 vs. 2.3 mg of CLA g-1 lipid, respectively).

The use of pastures in the diet of animals in comparison with the preserved forage, such as hay or silage, has significant effect on the concentration of CLA and has more advantages for pasture-based productive systems. In addition to minimizing production costs associated with preservation of forages and offering greater production stability, as well as reducing dependence on external factors and taking greater advantage of natural conditions of climate and soil, it also promotes the sustainability of the entire production system and, if implemented, will offer a product of better nutritional quality to the final costumer.

CONCLUSION

In any system of beef cattle production, especially pasture production, it is necessary to readjust nutritional plans that are in line with the performance enhancement of these animals. The choice of fodder, as well as appropriate management, are significant steps that become decisive to the productive performance results of the animals.

The appropriate management of the pasture involves sustainability and balance of nutrients for sufficient dry matter production, which will then be used in the feeding of the animals. The chemical elements present in the soil-plant-animal system have important functions that make it work normally. Thus, strategies that aim to increase the availability of dry matter of the forage directly lead to an increase in the basal resources and consequently, a decrease in the need to introduce additional resources to the system (Paulino et al., 2006). It makes the productive cycle more efficient and profitable, since tropical grasses are the cheapest sources of energy that can be converted into animal products for cattle.

From this perspective, we can affirm that the animal production in pastures results from the interaction of several factors inherent to the production of forage, consumption and conversion of the ingested fodder into the final product, considering that forage quality is basically defined by the potential the feed has in generating animal performance. Thus, research and management of the productive process, a study of the responses of plants and animals in specific conditions of environment, respecting their requirements and physiological rhythms, allowed reaching significant advances in green beef production.

REFERENCES

Andrade, C. M. S. and Assis, G. M. L. (2008). *Capim-Xaraés: cultivar de gramínea forrageira recomendada para pastagens no Acre.* [*Xaraés grass: cultivar of forage grass recommended for grazing in Acre*]. Embrapa Acre-Documentos (INFOTECA-E).

ANUALPEC, 2013. *Anuário da Pecuária Brasileira.* Instituto FNP: São Paulo. [ANUALPEC, 2013. Yearbook of Brazilian Livestock. FNP Institute: São Paulo].

Aronovich, S. (1995). O capim colonião e outros cultivares de *Panicum maximum* (Jacq.): Introdução e evolução do uso no Brasil. *Simpósio sobre Manejo da Pastagem* [Colonião grass and other cultivars of *Panicum maximum* (Jacq.): Introduction and evolution of use in Brazil. *Symposium on Pasture management*], 12, 1-20.

Balsalobre, M. A. A. (2002). *"Valor alimentar do capim tanzânia irrigado".* ["Feed value of irrigated Tanzania grass"]. Doctoral Thesis, Escola Superior de Agricultura Luiz de Queiroz, University of São Paulo, Piracicaba

Balsalobre, M. A., Nussio, L. G., Martha Jr, G. B. and Mattos, W. (2001). Controle de perdas na produção de silagens de gramíneas tropicais. *Reunião Anual da sociedade Brasileira de Zootecnia* [Loss control in the production of silage from tropical grasses. *Annual Meeting of the Brazilian Society of Animal Science*], 38, 890-911.

Barcellos, M. D. (2007) *"Beef lovers": Um estudo cross-cultural sobre o comportamento de compra da carne bovina.* 328 p. Tese (Doutorado em Agronegócios) – Programa de Pós-Graduação em Agronegócios ["Beef lovers": A cross-cultural study on beef purchasing behavior. 2007. 328 f. Thesis (PhD in Agribusiness) - Graduate Program in Agribusiness], Universidade Federal do Rio Grande do Sul, Porto Alegre.

Bauman, D. E. and Griinari, J. M. (2001). Regulation and nutritional manipulation of milk fat: low-fat milk syndrome. *Livestock Production Science,* 70(1), 15-29.

166 *Bruno Lala, Vinícius Valim Pereira, Ulysses Cecato et al.*

Bressan, M. C., Rossato, L. V., Rodrigues, E. C., Alves, S. P., Bessa, R. J. B., Ramos, E. M. and Gama, L. T. (2011). Genotype× environment interactions for fatty acid profiles in and finished on pasture or grain. *Journal of animal science*, 89(1), 221-232.

Bridi, A. M., Constantino, C., Tarsitano, M. A. (2011). Qualidade da carne de bovinos produzidos em pasto. In: *Simpósio de Produção Animal a Pasto* [Quality of beef produced in pasture. In: *Symposium of Animal production to Pasture*]. 311-332. ISBN 978-85-63633-10-1. Maringá. Sthampa.

Brondani, I. L. (2002) *Desempenho e características da carcaça de bovinos jovens* [*Performance and carcass characteristics of young bovines*]. UNESP – Universidade Estadual Paulista. Faculdade de Ciências Agrárias e Veterinárias. (Tese de Doutorado em Zootecnia) [College of Agrarian and Veterinary Sciences. (PhD in Animal Science)]. Jaboticabal, 133 p.

Carloto, M. N., Euclides, V. P. B., Montagner, D. B., Lempp, B., Difante, G. D. S. and Paula, C. C. L. D. (2011). Animal performance and sward characteristics of xaraés palisade grass pastures subjected to different grazing intensities, during rainy season. *Pesquisa Agropecuária Brasileira,* 46(1), 97-104.

Carvalho Júnior, A. M. D., Pereira Filho, J. M., Silva, R. D. M., Cezar, M. F., Silva, A. M. D. A. and Silva, A. L. N. D. (2009). Efeito da suplementação nas características de carcaça e dos componentes não-carcaça de caprinos F1 Boer×SRD terminados em pastagem nativa. *Revista Brasileira de Zootecnia* [Effect of supplemental feeding on carcass and non-carcass characteristics of F1 (Boer × SRD) goats finished on native pasture. *Brazilian Journal of Animal Science*]. 38(7), 1301-1308.

Costa, E. C. D., Restle, J., Brondani, I. L., Perottoni, J., Faturi, C. and Menezes, L. D. (2002). Composição física da carcaça, qualidade da carne e conteúdo de colesterol no músculo *Longissimus dorsi* de novilhos Red Angus superprecoces, terminados em confinamento e abatidos com diferentes pesos [Carcass Composition, Meat Quality and Cholesterol Content in the *Longissimus dorsi* Muscle of Young Red

The Brazilian Green Beef 167

Angus Steers Confined and Slaughtered with Different Weights]. *Revista Brasileira de Zootecnia,* 31(1), 417-428.

Da Silva, S. C. (2009). Conceitos básicos sobre sistemas de produção animal em pasto. Intensificação de Sistemas de Produção Animal em Pasto [Basic concepts on animal production systems in pasture. *Intensification of Animal production Systems in Pasture*], 25, 7-36. In: Da Silva, S. C.; Pedreira, C. G. S.; Moura, J. C. (Eds.). Intensificação de sistemas de produção animal em pasto [*Intensification of animal production systems in pasture*]. Piracicaba: FEALQ.

Da Silva, S. C. D. and Nascimento Júnior, D. D. (2007). Avanços na pesquisa com plantas forrageiras tropicais em pastagens: características morfofisiológicas e manejo do pastejo. *Revista Brasileira de Zootecnia* [Advances in the research with tropical forage plants in pastures: morphophysiological characteristics and grazing management. *Brazilian Journal of Animal Science*], 36, 122-138.

Di Marco, O. N. (1998). *Crescimiento de vacunos para carne* [*Growth of beef cattle*]. Mar Del Plata, Argentina. 246p.

Enser, M.; Hallet, K. G.; Hewett, B.; Fursey, G. A. J.; Wood, J. D.; Harrington, G. (1998). Fatty acid content and composition of UK beef and lamb muscle in relation to production system and implications for human nutrition. *Meat Science*, v.49, n.3, p.329-341.

Euclides, V. P. B., Macedo, M. C. M., Valle, C. D., Flores, R. and Oliveira, M. D. (2005). Animal performance and productivity of new ecotypes of *Brachiaria brizantha* in Brazil. In: *International Grassland Congress* (Vol. 20, p. 106).

Euclides, V. P. B. and de Medeiros, S. R. (2003). *Valor nutritivo das principais gramíneas cultivadas no Brasil. Embrapa Gado de Corte.* [*Nutritive value of the main grasses grown in Brazil. Embrapa Beef Cattle*].

Euclides, V. P. B., Macedo, M. C. M. and Oliveira, M. P. D. (1995a). Avaliação de ecotipos de *Panicum maximum* sob pastejo em pequenas parcelas. *Reunião Anual da Sociedade Brasileira de Zootecnia.* [Evaluation of *Panicum maximum* ecotypes under grazing in small plots. *Annual Meeting of the Brazilian Society of Animal Science*].

168 *Bruno Lala, Vinícius Valim Pereira, Ulysses Cecato et al.*

Euclides, V., Macedo, M. and Valle, L. (1995b). Avaliação de acessos de *Panicum* maximum sob pastejo. *Campo grande: Embrapa-CNPGC.* [Evaluation of *Panicum* maximum access under grazing. *Large field: Embrapa-CNPGC*].

Fonseca, D. M., Santos, M. E. R. and Martuscello, J. A. (2010). Importância das forrageiras no sistema de produção. *Plantas forrageiras. Viçosa: UFV,* 13-29. [*Importance of* forage *in the production system.* Forage *plants. Viçosa: UFV,* 13-29].

French, P., Stanton, C., Lawless, F., O'riordan, E. G., Monahan, F. J., Caffrey, P. J. and Moloney, A. P. (2000). Fatty acid composition, including conjugated linoleic acid, of intramuscular fat from steers offered grazed grass, grass silage, or concentrate-based diets. *Journal of Animal Science,* 78(11), 2849-2855.

Fugita, C. A., Prado, I. N., Jobim, C. C., Zawadzki, F., Valero, M. V., Pires, M. C. O., Prado, R. M. and Françozo, M. C. (2012). Corn silage with and without enzyme-bacteria inoculants on performance, carcass characteristics and meat quality in feedlot finished crossbred bulls. *Brazilian Journal of Animal Science,* 41(1), 154-163.

Galbeiro, S. *"Características morfogênicas, acúmulo e qualidade de forragem do capim-Xaraés submetido a intensidades de pastejo sob lotação contínua."* Tese (Doutorado em Zootecnia) – Universidade Estadual de Maringá, Maringá. 2009. 84p. ["Morphogenic characteristics, accumulation and forage quality of Xaraés grass subjected to grazing intensities under continuous stocking." Thesis (PhD in Animal Science) - State University of Maringá, Maringá. 2009. 84p].

Garcia, P. T., Pensel, N. A., Sancho, A. M., Latimori, N. J., Kloster, A. M., Amigone, M. A. and Casal, J. J. (2008). Beef lipids in relation to animal breed and nutrition in Argentina. *Meat Science,* 79(3), 500-508.

Gomide, C. A. M., Gomide, J. A. (2001) Morphogenesis and growth analysis of Mombaça grass in the establishment and aftermaths growths. In: *XIX Int. Grassland Congress Proceedings*, São Pedro-SP, 64-65.

Gonçalves, E. N., Carvalho, P. C. D. F., Kunrath, T. R., Carassai, I. J., Bremm, C. and Fischer, V. (2009). Relações planta-animal em ambiente pastoril heterogêneo: processo de ingestão de forragem. *Brazilian Journal of Animal Science* [Plant-animal relationships in a heterogeneous pastoral environment: forage intake process. *Brazilian Journal of Animal Science*]. Viçosa, MG. 38:9, 1655-1662.

Hankins, O. G. and Howe, P. E. (1946) *Estimation of the composition of beef carcasses and cuts.* Washington: USDA, 20p. (Technical Bulletin, 926).

Harfoot, C. G. (1981). Lipid metabolism in the rumen. In: Christie, W. W. (Ed.) *Lipid metabolism in ruminant animals.* Pergamon Press: Oxford, UK. 21–55.

Hayashi, A. A. (2003). *"Efeito da suplementação com ácido linoléico conjugado (CLA) na composição do leite, no perfil de ácidos graxos e na atividade de enzimas lipogênicas em ratas lactantes."* Dissertação de Mestrado, Escola Superior de Agricultura Luiz de Queiroz, Universidade de São Paulo, Piracicaba. [*"Effect of supplementation with conjugated linoleic acid (CLA) on milk composition, fatty acid profile and activity of lipogenic enzymes in lactating rats."* Master's Dissertation, Luiz de Queiroz College of Agriculture, University of São Paulo, Piracicaba].

Hopper, T. H. (1944). Methods of estimating the physical and chemical composition of cattle. *Journal of Agricultural Research, 68,* 239-268.

Igarasi, M. S., Arrigoni, M. D. B., Hadlich, J. C., Silveira, A. C., Martins, C. L. and Oliveira, H. N. D. (2008). Características de carcaça e parâmetros de qualidade de carne de bovinos jovens alimentados com grãos úmidos de milho ou sorgo. *Revista Brasileira de Zootecnia* [Carcass characteristics and meat quality parameters of young cattle fed with moist grains of corn or sorghum. *Brazilian Journal of Animal Science*], 37(3), 520-528.

Instituto Brasileiro De Geografia E Estatística (IBGE) (2017). *Estatística da Produção Pecuária.* Indicadores IBGE [*Livestock Production Statistics.* Indicators IBGE].

170 *Bruno Lala, Vinícius Valim Pereira, Ulysses Cecato et al.*

Instituto Brasileiro De Geografia E Estatística (IBGE) (2014). *Estatística da Produção Pecuária.* Indicadores IBGE. 2017 [*Livestock Production Statistics.* Indicators IBGE. 2017].

Ip, C. and Pariza, M. (2005) CLA (Conjugated linoleic acid). *Interpretative review of recent nutrition research.* Available on: www.nationaldairy council.org.

Jank, L. (1995). Melhoramento e seleção de variedades de *Panicum maximum. Simpósio sobre manejo da pastagem* [Improvement and selection of varieties of *Panicum maximum. Symposium on pasture management*], 12, 21-58.

Jarrige, R., Grenet, E., Demarquilly, C. and Besle, J. M. (1995). Les constituants de l'appareil végétatif des plantes fourragères. In: Jarrige, R.; Ruckebusch, Y.; Demarquilly, C.; Farcen, M. H.; Journet, M. (Ed.) *Nutrition des ruminants domestiques - Ingestion et digestion.* INRA: Paris, 25-81. [The constituents of the vegetative apparatus of forage plants. In: Jarrige, R.; Ruckebusch, Y.; Demarquilly, C.; Farcen, M. H.; Journet, M. (Ed.) *Nutrition of domestic ruminants - Ingestion and digestion.* INRA: Paris, 25-81].

Jorge A. M. and Fontes, C. A. A. (2001). Composição física da carcaça de bovinos e bubalinos abatidos em diferentes pesos. In: *Congresso Brasileiro De Ciência E Tecnologia De Carnes* [Physical composition of the carcass of cattle and buffaloes slaughtered at different weights. In: *Brazilian Congress of Science and Technology of Meat*], São Pedro, Campinas: Instituto de Tecnologia de Alimentos, 82-83.

Lucas, R. C. *"Efeito do genótipo sobre as características quantitativas e qualitativas da carcaça de caprinos terminados em pastagem nativa."* 2007. 65p. Dissertação (Mestrado em Zootecnia). [*"Effect of the genotype on the quantitative and qualitative characteristics of the carcass of goats finished in native pasture."* 2007. 65p. Dissertation (Master in Animal Science)] - Universidade Federal de Campina Grande, Patos, 2007.

Lush, J. L. (1926). Practical methods of estimating the proportions of fat and bone in cattle slaughtered in commercial packing plants. *Journal of Agricultural Research,* 32, 727-755.

Macedo, M. C. M. (1995). Pastagens no ecossistema Cerrados: pesquisas para o desenvolvimento sustentável. *Anais do 32º simpósio sobre pastagens nos ecossistemas brasileiros.* Sociedade Brasileira de Zootecnia, Brasília. [Pastures in the Cerrados ecosystem: research for sustainable development. *Proceedings of the 32nd Symposium on Pastures in Brazilian Ecosystems.* Brazilian Society of Animal Sciences, Brasília].

McGuire, M. A. and McGuire, M. K. (2000). Conjugated linoleic acid (CLA): A ruminant fatty acid with beneficial effects on human health. *Journal of Animal Science,* 77(E-Suppl), 1-8.

Medeiros, S. R., Oliveira, D. E., Aroeira, L. J. M., McGuire, M. A., Bauman, D. E. and Lanna, D. P. D. (2010). Effects of dietary supplementation of rumen-protected conjugated linoleic acid to grazing cows in early lactation. *Journal of Dairy Science,* 93(3), 1126-1137.

Mertens, D. R. (1994). *Regulation of forage intake. Forage quality, evaluation, and utilization, (foragequalityev)*, 450-493.

Missio, R. L., Restle, J., Moletta, J. L., Kuss, F., Neiva, J. N. M. and Moura, I. C. (2013). Características da carcaça de vacas de descarte abatidas com diferentes pesos. *Revista Ciência Agronômica* [Characteristics of livestock cattle carcasses when slaughtered at different weights. *Revista Ciência Agronômica*], 44(3), 644.

Pariza, M. W., Park, Y. and Cook, M. E. (2001). The biologically active isomers of conjugated linoleic acid. *Progress in Lipid Research,* 40(4), 283-298.

Paulino, M. F., Zamperlini, B., Figueiredo, D. M., Moraes, E. H. B. K., Fernandes, H. J., Porto, M. O., Sales, M. F. L., Paixão, M. L., Acedo, T. S., Detmann, E. and Valadares Filho, S. C. (2006). Bovinocultura de precisão em pastagens. *Simpósio De Produção De Gado De Corte.* [Cattle breeding precision in pastures. *Beef Cattle Production Symposium*], 5, 361-412.

Penati, M. A. (2002). *"Estudo do desempenho animal e produção do capim Tanzânia (Panicum maximum, Jacq.) em um sistema rotacionado de pastejo sob irrigação em três níveis de resíduo pós-*

172 *Bruno Lala, Vinícius Valim Pereira, Ulysses Cecato et al.*

pastejo" ["Study of animal performance and yield of Tanzania grass (*Panicum* maximum, Jacq.) In a rotational grazing system under irrigation at three levels of post-grazing residue"] (Doctoral dissertation, University of São Paulo).

Ponnampalam, E. N., A. J. SinCLAir, A. R. Egan, S. J. Blakeley, B. J. Leury. 2001. Effect of diets containing n−3 fatty acids on muscle long-chain n−3 fatty acid content in lambs fed low- and medium-quality roughage diets. *Journal of Animal Science* 79:698–706.

Realini, C. E., Duckett, S. K., Brito, G. W., Dalla Rizza, M. and De Mattos, D. (2004). Effect of pasture vs. concentrate feeding with or without antioxidants on carcass characteristics, fatty acid composition, and quality of Uruguayan beef. *Meat Science,* 66(3), 567-577.

Reis, R. A. R. (1993). *Valor nutritivo de plantas forrageiras* [*Nutritive value of forage plants*]. FCAVJ-UNESP: FUNEP.

Restle, J., Roso, C., Aita, V., Nornberg, J. L., Brondani, I. L., Cerdótes, L. and Carrilo, C. (2002). Produção animal em pastagem com gramíneas de estação quente. *Revista Brasileira de Zootecnia* [Animal performance in Summer Grasses Pastures. *Brazilian Journal of Animal Sciences*], 31(3), 1491-1500.

Saab, M. S. B. L. M. (1999) *"Valor percebido pelo consumidor: um estudo de atributos da carne bovina."* 154p. Dissertação (Mestrado em Administração) – Programa de Pós-Graduação em Administração. Faculdade de Economia, Administração e Contabilidade. Universidade de São Paulo. São Paulo. [*"Perceived value by the consumer: a study of beef attributes."* 154p. Dissertation (Master in Administration) - Post-Graduate Program in Administration. College of Economics, Administration and Accounting. University of Sao Paulo. São Paulo].

Savidan, Y. H., Jank, L. and Costa, J. C. G. (1990). Registro de 25 acessos selecionados de *Panicum maximum. Embrapa Gado de Corte-Documentos (INFOTECA-E).* [Record of 25 selected hits of *Panicum maximum. Embrapa Beef Cattle Documents (INFOTECA-E)*].

Spers, E. E. (2003) *"Mecanismos de regulação da qualidade e segurança em alimentos."* 136 p. Tese (Doutorado em Administração) – Programa de Pós-Graduação em Administração, Faculdade de

Economia, Administração e Contabilidade, Universidade de São Paulo. São Paulo. [*"Mechanisms for regulation of food quality and safety."* 136 p. Thesis (PhD in Administration) - College of Economics, Administration and Accounting, University of São Paulo. São Paulo].

Strydom, P. E., Naude, R. T., Smith, M. F., Scholtz, M. M. and Van Wyk, J. B. (2000). Characterization of indigenous African cattle breeds in relation to carcass characteristics. *Animal Science*, 70(2), 241-252.

Thiago, L. And Gill, M. (1990). *Consumo voluntário de forragens por ruminantes: mecanismo físico ou fisiológico? Bovinocultura de corte.* [*Voluntary feed intake by ruminants: physical or physiological mechanism? Beef Cattle*].

Todaro, M., Corrao, A., Alicata, M. L., Schinelli, R., Giaccone, P. and Priolo, A. (2004). Effects of litter size and sex on meat quality traits of kid meat. *Small Ruminant Research*, 54(3), 191-196.

United States Department of Agriculture (USDA). Agricultural Research Service. Brazil. Livestock and Products Annual. *Annual Livestock* 2016. Gain Report Number: BR1614. 2016.

Valle, C. D., Euclides, V. P. B. and Macedo, M. C. M. (2000). Características das plantas forrageiras do gênero *Brachiaria*. *Simpósio Sobre Manejo Da Pastagem* [Characteristics of forage plants of the genus *Brachiaria*. *Symposium on Pasture Management*], 17, 65-108.

Van Soest, P. J. (1994). *Nutritional ecology of the ruminant*. 2 ed. Ithaca: Cornell. 476p.

Williams, C. M. (2000). Dietary fatty acids and human health. In Annales de Zootechnie [Dietary fatty acids and human health. In *Animal Science Reports*] 49:3, 165-180.

World Health Organization (WHO). Interim summary of conclusions and dietary recommendations on total fat & fatty acids. *Report of a joint WHO/FAO expert consultation*. Geneva. 2008.

Zawadzki, F., do Prado, I. N., and Prache, S. (2013). Influence of level of barley supplementation on plasma carotenoid content and fat spectrocolorimetric characteristics in lambs fed a carotenoid-rich diet. *Meat Science*, 94(3), 297-303.

In: Beef: Production and Management Practices ISBN: 978-1-53613-254-0
Editor: Nelson Roberto Furquim © 2018 Nova Science Publishers, Inc.

Chapter 7

NUTRITIONAL AND GENETIC FACTORS THAT AFFECT MEAT QUALITY

Otávio Rodrigues Machado Neto[1,], PhD*
Josiane Fonseca Lage[2], PhD, Liziana Maria Rodrigues[3]
and Mateus Silva Ferreira[4]

[1]Department of Animal Production – Faculdade de Medicina
Veterinária e Zootecnia/UNESP – Botucatu, SP - Brasil
[2]Trouw Nutrition Brasil, Campinas, SP - Brasil
[3]Animal Science – Departamento de Zootecnica/UFLA - Lavras,
MG - Brasil
[4]Animal Science – Faculdade de Ciências Agrárias e
Veterinárias/UNESP – Jaboticabal, SP - Brasil

1. PRENATAL NUTRITION AND MEAT QUALITY

Meat production for human consumption depends on the amount of muscle produced during animal growth. The muscle development starts during the fetal phase. After birth, the number of fibers does not change

---

[*] Corresponding Author Email: otaviomachado@fmvz.unesp.br.

and there is only increase in fiber size and in muscle mass. According to Gionbelli, (2015), beef cattle production system has potential to increase about 30 to 40% on nutrition, genetic and reproduction with improvements on the breeding phase.

Fetal or prenatal development is a complex biological event influenced by several genetic, epigenetic and maternal factors capable of changing the size and functional capacity of placenta, utero-placental blood flow, nutrients and oxygen transfer from the mother to the fetus, availability of nutrients for conception, endocrine interactions and metabolic routes (Wu et al., 2006).

Fetal programming mechanism may occur during cell division in response to a recent stimulus and transferred to other cells (Bonasio; TU; Reinberg, 2010). Maternal malnutrition can cause adaptation in offspring leading to metabolic changes to save energy such as increase in fat deposition and decrease in muscle mass in the progeny's body (Blair et al., 2013).

Carcass characteristics and meat quality in cattle depend on the composition and proportion of muscle, adipose and connective tissues in the skeletal muscle, as well as the number and size of muscle fibers.

During gestation, the skeletal muscle presents lower priority in the allocation of nutrients over vital organs and tissues such as the brain and the cardiovascular system. Therefore, skeletal musche development may be more vulnerable to nutritional variations (Zhu et al., 2006). Thus, there may be an effect on muscle growth throughout the animal life. Figure 1 presents the stages of development and growth of skeletal muscle and adipose tissue of beef cattle and the possible effects of maternal nutrition in each phase.

Connective tissue presents little participation in mass composition. However, a high proportion of this type of tissue can reduce meat tenderness (Wu et al., 2006). Regarding meat production, a high number of muscle fibers can optimize lean meat percentage and growth efficiency. Muscle hyperplasia, which is the increase in muscle fibers, is fixed and is normally defined until the end of gestation. However, the increase in size of muscular fiber (hypertrophy) can be affected by several factors after birth (Rehfeldt et. al., 2004).

Therefore, the best way to increase meat production without damaging its quality is selecting animals with a greater number of moderate diameter muscle fibers (Lee et al., 2010; Rehfeldt et. al., 2004). Nutrient deficiency during gestation might reduce the number of muscle fibers, muscle mass and, consequently, affect meat quality and progeny's characteristics.

According to Blair et al., (2013) there was a trend for fat thickness reduction and inferior classification at the Quality Grade in cattle whose mothers presented poor body condition and energy status during the middle quarter of the gestation. The mother's energetic *status* during the middle third of gestation plays an important role in the development of progeny's carcass characteristics (Blair et al., 2013).

Greenwood et al., (2005) observed that maternal malnutrition may represent a reduction of up to 35% in progeny birth weight. Besides, animals with lower birth weight have limited capacity for compensatory growth.

Then, when animals are slaughtered at the same age (30 months), there are significant differences in carcass weight and meat yield with better performance in animals that had a higher birth weight. However, when animals are slaughtered according to their body weight, there is no difference in carcass characteristics since heavier animals entered the feedlot earlier and were slaughtered younger.

Meat yield and quality may be affected by many muscle biological properties such as myofibers number, size, composition or types (Bi; Kuang, 2012). The skeletal muscle expresses different isoforms of Myosin gene (MyHC), which defines the type of muscle fiber. These isoforms are translated and differentiated according to animal age and muscle location (Zhang et al., 2014) and differ in metabolic properties producing fibers of slow contraction I oxidative type (MyHC-I), quick contraction glycolytic-oxidative type IIa (MyHC-IIa), quick contraction glycolytic type IIb (MyHC-IIb) and type IIx (MyHC-IIx) (Schiaffino; Reggiani, 1996).

According to Zhang (2014), younger bulls present higher quantity of type IIa muscle fibers, which leads to a more flexible metabolic properties for growth since this type of fiber has a fast-oxidative-glycolytic intermediate metabolism. On the other hand, bovine muscle could be adapted to slow oxidation conditions because there is not Type IIb fiber in *Longissimus dorsi*, *Semitendinosus* and *Soleus*; and this type of fiber is the most glycolytic among MyHCs.

The composition of skeletal muscle also presents great importance in meat quality. Since the relation of the type of oxidative and glycolytic fibers can be affected by the diet (Vestergaard et al., 2000), concentrations of intramuscular glycogen and the rate of glycolysis after slaughter affect the production of lactic acid, meat pH and water retention capacity. Likewise, the increase in intramuscular fat promotes post-slaughter lipid peroxidation resulting in the oxidation of muscle components such as oximioglobin, changing meat color and flavor (Wu et al., 2006).

The amount of intramuscular fat is a fundamental characteristic for meat palatability. Furthermore, it is responsible for marbling which contributes to meat juiciness and flavor (Du et al., 2010). The ratio between tissue types and muscle composition depends on the cellular differentiation during the fetal phase. Muscle and adipose tissues are originated by mesenchymal stromal cells (MSC) through the action of transcription factors (TF) that regulate the commitment and differentiation of these cells. The TF can act in different ways depending on the concentrations, cell-to-cell interactions and extra-cellular matrix (Ladeira et al., 2016).

The determining factor for the type of tissue to be formed occurs through the signaling *Wnt* (*Wingless and Int*). The signaling *Wnt* occurs in an autocrine and paracrine way, inducing the proliferation, differentiation or maintenance of cellular precursors (Du et al., 2010).

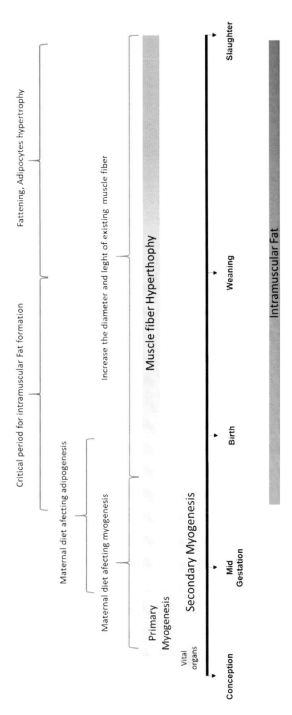

Source: Adapted from Du et al., (2010) and Du; Ford; Zhu (2017).

Figure 1. Development of skeletal muscle and adipose tissue in beef cattle and effect of maternal nutrition on hyperplasia and hypertrophy of these tissues (approximate times).

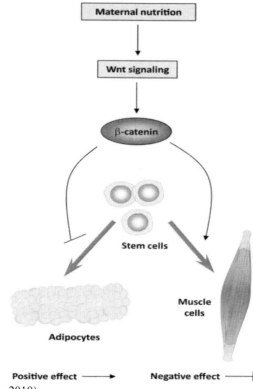

Source: (Du et al., 2010).

Figure 2. *Wnt* signaling and development of fetal skeletal muscle. *Upregulation* of the *Wnt signaling* promotes myogenesis while downregulation promotes adipogenesis.

Wnt's pathway signaling depends on β-Catenin signaling (Figure 2). The activation of Wnt signal inhibits "β-Catenin destruction complex" composed of *axin*, GSK-3β (*glycogen synthesis kinase-3 β*) and anaphase promoter complex, accumulating cytoplasmic β-Catenin. Part of this cytoplasmic β-Catenin enters the cell nucleus and interacts with T cells transcription factors by activating specific genes. Then, β-Catenin is able to regulate the growth of embryonic, prenatal, postnatal and oncogenic tissues.

In the skeletal muscle, β-Catenin regulates the expression of the transcription factor Pax3 (*Paired box* 3) and Gli. Pax3 stimulates myogenesis by acting against MyoD while Gli plays an important role in the expression of MyF5. Therefore, the expression of MRFs (*myogenic regulatory factors*) is controlled by *Wnt* signaling and *Sonic hefgefog* via Pax3 and Pax7, blocking the β-Catenin pathway and reducing the total number of myocytes. Likewise, adipogenesis is also controlled by *Wnt* signaling. The activity of PPARg is also regulated by GSK-3β which allows β-Catenin to escape from proteosomal degradation, leading to the inhibition of PPARg expression in targeted genes. (Du et al., 2010).

So, *Wnt* signaling increases myogenesis and reduces adipogenesis in the skeletal muscle, participating in the regulation of body fat reducing and susceptibility to obesity.

Considering fetal myogenesis and adipogenesis (Figure 1), it is possible to manipulate maternal diet during gestation to increase the expression of *Wnt* signaling in the beginning and middle of gestation. Myogenesis occurs and inhibits *Wnt* signaling at the end of gestation favoring the adipogenesis in this period. (Du et al., 2010).

Another way of adipocyte formation is through non-myogenic progenitors, also known as fibro-adipogenic precursors (FAP). Intramuscular fat is considered a connective tissue whose development is associated with fibrogenesis. FAP has adipogenic, fibrogenic osteogenic and chondrogenic potential. They act in the recovery of muscle damage and are also responsible for the development of intramuscular fat (Ladeira et al., 2016). The end of gestation is the period of greater production of intramuscular adipocytes and the manipulation of maternal diet for greater expression of FAP can lead to greater marbling and meat quality.

After formation, adypocytes hypertrophy occurs through the accumulation and storage of fat during growth. *Peroxisome proliferator-activated receptor* (PPAR) are nuclear receptors that bind to fatty acids and act on the regulation of nutrients and energy homeostasis (Ladeira et al., 2016). These FT connect to binding proteins (FABP and LPL, for example) regulating the storage of fatty acids and adipogenesis. According to Bispham et al., (2005), PPARa and PPARg use fatty acids as endogenous binders, suggesting that these transcription factors can be regulated by the diet. However, due to the low concentration of free fatty acids in cattle fetus, this mechanism may be limited. These authors observed that maternal nutritional restrictions up to 80 days of gestation increased the expression of PPARg in fetus and presented a higher amount of adipose tissue in relation to fetuses whose mothers received *ad libitum* diets (Bispham et al., 2005).

Furthermore, the fetal phase is considered the best moment for the manipulation of diet aiming at the increase of marbling due to the abundance of multipotent cells and the provision of readily available nutrients coming from the maternal diet. There is still formation of adipose cells after birth and up to 250 days of life. After this period, the efficiency of dietary manipulation is conditioned to hypertrophy of existing adipocytes (Du et al., 2010).

Duarte et al., (2014), comparing the gene expression of skeletal muscle components of fetuses coming from overnourished pregnant cows or control, observed an increase in gene expression favorable to fibrogenesis and adipogenesis in the fetus of undernourished cows without affecting the myogenesis of these animals. This may be related to the fact that control cows do not present enough restriction once their maintenance requirements have been met.

Du et al., (2004) observed that in pregnant cows with 50% nutritional restriction at the beginning of gestation, the levels of Calpainas I and II in maternal and fetal skeletal muscle were not affected. However, this restriction reduced the concentration of calpastatin (a specific inhibitor of calpain) in maternal and fetal muscle. Nutritional restriction also reduced phosphorylated mTOR concentrations in both the mother's skeletal muscle the fetus' and increased concentrations of ubiquitinated proteins in the maternal muscle.

Growth factors, such as the IGF system, also influence the proliferation and differentiation of cells in the process of myogenesis. IGF-I and IGF-II are growth factors with endocrine, autocrine and paracrine action, respectively, and have several effects including erythropoiesis stimulation, anabolism, cell growth, cell differentiation and satellite cell proliferation. IGF-I is the major regulator of myogenesis, stimulating differentiation by the induction of myogenin expression. The nutritional status is an important systemic modulator of IGF-I before and after birth. In fasting animals and undernourished matrixes, IGF-I and hepatic

Nutritional and Genetic Factors That Affect Meat Quality 183

production of IGF-I are reduced due to the reduction of GH receptors. Consequently, it reduces the number of receptors for IGF-I during starvation. IGF-II is also reduced with fetal malnutrition (Rehfeldt et al., 2004).

Table 1. Effects of maternal nutrition on progeny carcass features

Results/Observations	Reference
Higher marbling score for animals whose mothers were supplemented during gestation.	Larson et al., (2009)
Increased progeny tenderness of cows maintained on improved pasture and increased moisture in the carcass of animals whose mothers were raised on native pasture.	Underwood et al., (2010)
Animals whose mothers had a positive energetic balance during gestation presented greater thickness of subcutaneous fat and better classification in *Yield Grade*.	Blair et al., (2013)
Control diet cows (without supplementation) generated animals with higher empty body fat, greater subcutaneous fat thickness, greater tenderness, higher marbling score and, consequently, better classification in *Yield Grade*.	Summers et. Al. (2015)
Progeny of cows that received 70% of nutritional requirements during gestation had greater diameter of subcutaneous adipocytes and better *Yield Grade*.	Long et al., (2015)
Increased amount of intramuscular fat in relation to subcutaneous fat in animals whose mothers were in negative energetic balance during gestation.	Mohrhauser et al., (2015)
Cows receiving 129% of protein nutritional requirements produced calves with higher subcutaneous fat and better *Yield Grade* at slaughtering. These animals had even lesser quantity of insulin after feeding in relation to the progeny of the cows that received 100% of protein requirements.	Wilson et al., (2016)

Maternal nutrition during the beginning and middle of gestation can affect glucose metabolism and glucose exchange, causing sensitivity to IGF-1 (Du et al., 2015). On the other hand, changes in maternal nutrition can affect the proportion of muscle, adipocytes and fibroblast cells since they are originated from mesodermal cells and their processes compete during development.

Therefore, the restriction of nutrients in early pregnancy may induce sensitivity to IGF1. Gonzalez et al., (2013) observed a reduction in the concentrations of IGF-1 and IGF-2 in calves whose mothers received a diet with nutritional restriction, reducing the proliferation of myogenic cells and, consequently, negatively interfering in the formation of muscle fibers. However, once the nutrient intake is reestablished, these animals start to produce more adiposity, but the damage in muscle development cannot be reestablished, leading to obese animals.

Table 1 presents a compilation of data on the effect of maternal nutrition on progeny carcass treats. Differences in diets, breed, period and duration of nutritional restriction or supplementation of pregnant cows explain the differences observed in these experiments.

There are few studies on how fetal programming affects meat quality. Many signaling ways and transcription factors are affected by maternal diet during the development and growth of beef cattle. However, most studies use animals from temperate climate and little is known about the effects of maternal diet in tropical climate. Thus, food management must be adopted in pregnant cows to meet the requirements, avoiding loss in the body condition score during gestation.

2. POSTNATAL NUTRITION AND MEAT QUALITY

The increase in concentrate level in the diet and energy intake can bring benefits in meat quality such as meat tenderness, increase in the deposition of backfat and intramuscular fat.

Meat quality is one of the main problems of the industry since standardization, lack of finishing and inadequate tenderness are big problems found in the Brazilian production.

The deposition of backfat in the carcass is related to gender, genetic group, body weight, weight gain rate, maturity, but is also influenced by energy density of the diet. The slaughtering industry has given great

Nutritional and Genetic Factors That Affect Meat Quality 185

importance to deposition of subcutaneous fat since it works as a thermal insulation.

The backfat thickness plays an important role in the reduction of "cold shortening" during carcass cooling process (Dolezal et al., 1982), which is as a fast drop in muscle temperature (lower than 19-14°C) before reaching *rigor mortis*. When the muscle temperature is reduced (15-0°C), the sarcoplasmic reticulum does not work effectively and becomes unable to bind to calcium, accumulating calcium in the sarcoplasm. Due to the presence of ATP in the muscles, they contract at a maximum level, making the filaments slide one over the other, eliminating band I (i) of the sarcomere (Savell et al., 2005). This fact can reduce meat tenderness.

In general, Brazilian slaughterhouses prefer carcasses with a minimum subcutaneous fat thickness of 3 mm. Below this, there is darkening of the external part of the muscles exposed to cooling process, giving an undesirable appearance. Over 6 mm thickness, it is a disadvantage to the the producer due to fat excess trimming before weighing and, for the slaughterhouses, it represents higher operational costs (Costa et al., 2002). So, the increase in backfat thickness improves meat tenderness by allowing the carcass to cool slowly and increase the enzymatic activity (Smith et al., 1976), resulting in better meat quality.

Consumers require beef with low contents of total lipids, saturated fatty acids and calories and high levels of polyunsaturated fatty acids which are important in the prevention of cardiovascular diseases. However, besides being an energy source, fat is also an important source of essential fatty acids and a carrier of fat soluble vitamins. In addition, it provides flavor and aroma to beef (Luchiari Filho, 2000).

The deposition of fat in the carcass is highly heritable and the total fat content may differ significantly among breeds (Itoh et al., 1999). The growth stage also influences on the total fat of the carcass, with certain breeds accumulating more fat in the growth phase in relation to others (DeSmet et al., 2004).

The accumulation of backfat is apparently regulated by different factors than those that regulate intramuscular fat accumulation. Smith and Crouse (1984) demonstrated that intramuscular adipocyte uses glucose and the subcutaneous adipocyte uses acetate as the primary substrate for the synthesis of fatty acids. Diets with starch promote greater accumulation of intramuscular fat than subcutaneous (Choat et al., 2003).

Intramuscular adipocytes are larger than subcutaneous tissue cells and present higher activity of hexokinase and phosphofructokinase. On the other hand, subcutaneous adipose tissue has higher levels of lipogenic enzymes such as NADP-malate dehydrogenase, phosphogluconate-6-dehydrogenase and glucose-6-phosphate dehydrogenase, evidencing exclusive roles in lipid metabolism (Miller et al., 1991; May et al., 1994).

During differentiation and in the selective expression of certain genes, the pre-adipocytes undergo morphological changes. The sequential expression of certain transcription factors such as SREBP and PPAR have a key role in the adipocyte conversion stages (Gregorie et al., 1998). SREBPs are transcription factors that play a central role in energy homeostasis, promoting lipogenesis and adipogenesis (Eberle et al., 2004).

Therefore, differences in the level of expression of SREBP gene may lead to differences in fatty acid composition in the adipose tissue of animals. Then, the interactions among nutrients in the diet and the expression of genes involved in lipid metabolism can illustrate innumerable possibilities regarding the deposition of fatty acids in the adipose tissue (Oliveira, 2013).

Lipoprotein lipase (LPL) is responsible for the breakdown of triacylglycerols from the circulation (lymph) in fatty acids and glycerol at the endothelial level. It is synthesized by adipocytes and acts as a cellular signal since it stimulates a new hyperplastic wave of adipose tissue. After this new proliferation wave, new cell signaling such as glycerol-3-phosphate dehydrogenase and fatty acid synthase are detected. Then, the cells start lipids accumulation, becoming adipocytes (Paulino et al., 2007).

Lipogenic enzymes acetyl CoA carboxylase (ACC) and fatty acid synthase are associated with the synthesis of lipids. In pigs and ruminants, it occurs in the adipose tissue; in birds and in humans, it occurs in the liver; and in rodents, it occurs in both places (Smith et al., 2003). Therefore, these enzymes are regulatory since changes in their activities reflect alterations in the fatty acid synthesis. Underwood et al., (2007) reported that cattle with greater amount of marbling fat had a lower rate of inactivation of the ACC gene.

About 46% of the changes in meat tenderness are related to animal genetics and 54% are explained by the environment (nutrition, management). When compared to animals of the same breed, genetic factors influence only 30% of the changes and 70% are controlled by environmental factors, showing a more restricted genetic influence (Koohmaraie 1995). Thus, meat tenderness does not depend only on the genotype of the animal but also on the association between genetics and environment. Several studies show that the increase in the proportion of *Bos Indicus* has proportional relation with stiffness (Whipple et al., 1990). Some authors associate zebu genes with the lack of meat tenderness, showing that animals crossbred with up to 25% of *Bos indicus* present similar characteristics in meat quality when compared to *Bos Taurus* (Wheller et al., 1994). Zebu animals have lower levels of calpain and higher levels of calpastatin, justifying the effect of these enzymes on meat tenderness (Wheller et al., 1990). However, some authors correlate meat tenderness only with calpastatin levels in the muscle, which is higher in zebu animals (Shackeford et al., 1991).

Besides, European animals accumulate subcutaneous fat at an inferior age than in zebu animals, protecting the carcasses of cold-shortening of muscle fibers (sarcomere) and reduction of muscle collagen. According to Crouse (1989), the main factors that differentiate tenderness among breeds are the lower myofibrillar fragmentation, myofibrillar fragmentation index and increased amount of connective tissue in the zebu. These animals also present higher concentrations of muscular calpastatin which is responsible for inhibiting the action of calpain in the meat tenderness.

According to Thompson's survey (1998) in MAS (Meat Australian Standard), meat from crossbred animals with 75% or more of zebu genotype were rejected in the palatability test (63% rejection by consumers), while animals with less than 25% of this genotype showed a disapproval of only 11%.

Ibrahim et al., (2008) identified higher amounts of calpastatin and lower amounts of calpain in *Longíssimus dorsi* of Brahma cattle, which

188   *O. Rodrigues Machado Neto, J. Fonseca Lage et al.*

justifies the meat tenderness of 4,27 kgf, which is low when compared to Waguli animals (Wagyu x Tuli) (3,48 kgf).

Lage et al., (2012), studying the effects of genetic groups (Nelore, Nelore x Angus and Nelore x Simental) on the quality of meat, found higher values for shear-force in Nelore when compared to the crossbred ones. The authors justified that this effect is due to the higher concentrations of calpastatin in *Bos Indicus*.

FINAL CONSIDERATIONS

Different phases of beef cattle production have an important effect on the characteristics of the final product. The Brazilian production chain needs improvements such as use of feedlots and high energy diets in the finishing stage. Therefore, the use of nutrition, technology and management strategies may increase intramuscular fat content in beef, contributing to the increase in beef added value.

REFERENCES

Bi, P.; Kuang, S. Meat Science and Muscle Biology Symposium: Stem cell niche and postnatal muscle growth. *Journal of Animal Science*, v. 90, n. 3, p. 924–935, 2012.

Bispham, J. et al., Maternal nutritional programming of fetal adipose tissue development: differential effects on messenger ribonucleic acid abundance for uncoupling proteins and peroxisome proliferator-activated and prolactin receptors. *Endocrinology*, v. 146, n. 9, p. 3943–3949, 2005.

Blair, A. D. et al., Pregnant Cow Nutrition : Effect on Progeny Carcass and Meat Characteristics. *The Range Beef Cow Simposium XXIII*, p. 41–50, 2013a.

Nutritional and Genetic Factors That Affect Meat Quality 189

Blair, A. D. et al., Pregnant cow nutrition: effect on progeny carcass and meat characteristics. *The Range Beef Cow Simposium XXIII.* Anais... Rapid City, South Dakota: 318, 2013b.

Bonasio, R.; Tu, S.; Reinberg, D. Molecular Signal of Epigenetic States. *Science*, v. 330, n. 6004, p. 612–616, 2010.

Du, M. et al., Effect of nutrient restriction on calpain and calpastatin content of skeletal muscle from cows and fetuses. *Journal of Animal Science,* v. 82, n. 9, p. 2541–2547, 2004.

Du, M. et al., Fetal programming of skeletal muscle development in ruminant animals. *Journal of Animal Science*, v. 88, 2010a.

Du, M. et al., Fetal programming in meat production. *Meat Science*, v. 109, p. 40–47, 2015.

Du, M.; Ford, S. P.; Zhu, M.-J. Optimizing livestock production efficiency through maternal nutritional management and fetal developmental programming. *Animal Frontiers*, v. 7, n. 3, p. 5, 2017.

Duarte, M. S. et al., Maternal overnutrition enhances mRNA expression of adipogenic markers and collagen deposition in skeletal muscle of beef cattle fetuses. *Journal of Animal Science*, 2014-7568-, 2014.

Gionbelli, M. P. Nutritional requirements for pregnant and non-pregnant beef cows. In: Valadares Filho, S. C., Costa E Silva, L. F., Lopes, S. A. et al., BR-Corte 3.0. *Cálculo de exigências nutricionais, formulação de dietas e predição de desempenho de zebuínos puros e cruzados* [*Calculation of nutritional requirements, formulation of diets and prediction of performance of pure and crossbred zebu*]. 2016. Cap. 13, pag 251-273. Available in: www.brcorte.com.br. Acesso em 13/09/2017.

Gonzalez, J. M. et al., Realimentation of nutrient restricted pregnant beef cows supports compensatory fetal muscle growth. *Journal of Animal Science,* v. 91, n. 10, p. 4797–4806, 2013.

Greenwood, P. L. et al., Consequences of nutrition and growth retardation early in life for growth and composition of cattle and eating quality of beef. *Journal of Animal Science*, v. 15, p. 183–196, 2005.

Ibrahim, R. M.; Goll, D. E.; Marchello, J. A.; Duff, G. C.; Thompsom, V. F.; Mares, S. W.; Ahmad, H. A. Effect of two dietary concentrate levels on tenderness, calpain, calpastatin activities, and carcass merit in Waguli and Brahman steers. *Journal of Animal Science*, v. 86, p. 1426-1433, 2008.

Koohmaraie, M.; Kent, M. P.; Shackelford, S. D.; Veiseth, E.; Wheeler, T. L. Meat tenderness and muscle growth: is there any relationship. *Meat Science*, v. 62, p. 345-352, 2002.

Ladeira, M. M. et al., Nutrigenomics and beef quality: A review about lipogenesis. *International Journal of Molecular Sciences*, v. 17, n. 6, 2016.

Larson, D. M. et al., Winter grazing system and supplementation during late gestation influence performance of beef cows and steer progeny. *Journal of Animal Science*, v. 87, n. 3, p. 1147–1155, 2009.

Lee, S. H.; Joo, S. T.; Ryu, Y. C. Skeletal muscle fiber type and myofibrillar proteins in relation to meat quality. *Meat Science*, v. 86, n. 1, p. 166–170, 2010.

Long, N. M. et al., Effects of early- to mid-gestational undernutrition with or without protein supplementation on offspring growth, carcass characteristics, and adipocyte size in beef cattle. *Journal of Animal Science*, v. 90, n. 1, p. 197–206, 2015.

Mohrhauser, D. A. et al., The influence of maternal energy restriction during mid-gestation on beef offspring growth performance and carcass characteristics. *Journal of Animal Science*, v. 93, n. 1, p. 786–793, 2015.

Rehfeldt, C., Fiedler, I., Stickland, N. C. Number and Size of Muscle Fibres in Relation to Meat Production. In: PAS, M. F. W.; Everts, M. E.; Haagsman, H. A. *Muscle development of livestock animals Physiology, Genetics and Meat Quality*. Cambridge, MA: CABI Publishing. p 2-30. 2004.

Savell, J. W.; Mueller, S. L.; Baird, B. E. The chilling of carcasses. *Meat Science*, v.70, p.449-459, 2005.

Schiaffino, S., Reggiani, C. Molecular diversity of myofibrillar proteins: gene regulation and functional significance. *Physiological Reviews*, vol. 76 n. 2, 371-423, 1996.

Shackelford, S. D.; Koohmaraie, M.; Miller, M. F.; Crouse, J. D.; Reagan, J. O. An evaluation of tenderness of the longissimus muscle of Angus by Hereford versus Brahman crossbred heifers. *Journal of Animal Science*, v. 69, n. 1, p. 171-177, 1991.

Summers, A. F.; Blair, A. D.; Funston, R. N. Impact of supplemental protein source offered to primiparous heifers during gestation on II. Progeny performance and carcass characteristics. *Journal of Animal Science*, v. 93 p. 1871-1880, 2015.

Underwood, K. R. et al., Nutrition during mid to late gestation affects growth, adipose tissue deposition, and tenderness in cross-bred beef steers. *Meat Science*, v. 86, p. 588–593, 2010.

Vestergaard, M. et al., Influence of feeding intensity, grazing and finishing feeding on muscle fibre characteristics and meat colour of semitendinosus, longissimus dorsi and supraspinatus muscle of young bulls. *Meat Science*, v. 54, p. 177–185, 2000.

Wheeler, T. L.; Shackelford, S. D.; Koohmaraie, M.. Variation in proteolysis, sarcomere length, collagen content, and tenderness among major pork muscles. *Journal of Animal Science*, v. 78, p. 958-965, 2000.

Whipple, G., Koomaraie, M., Dikeman, M. E., et al., Evaluation of attributes that affect longissimus muscle tenderness in Bos taurus and Bos Indicus cattle. *Journal of Animal Science*, v. 68. p. 2716-2728. 1990.

Wilson, T. B. et al., Influence of excessive dietary protein intake during late gestation on drylot beef cow performance and progeny growth, carcass characteristics, and plasma glucose and insulin concentrations. *Journal of Animal Science*, v. 94, n. 5, p. 2035–2046, 2016.

Wu, G. et al., Board-invited review: Intrauterine growth retardation: Implications for the animal sciences. *Journal of Animal Science*, v. 84, p. 2316–2337, 2006.

Zhang, M. et al., Expression of MyHC genes, composition of muscle fiber type and their association with intramuscular fat, tenderness in skeletal muscle of Simmental hybrids. *Molecular Biology Reports*, v. 41, n. 2, p. 833–840, 2014.

Zhu, M. J. et al., Maternal nutrient restriction affects properties of skeletal muscle in offspring. *The Journal of physiology*, v. 575, p. 241–250, 2006.

In: Beef: Production and Management Practices ISBN: 978-1-53613-254-0
Editor: Nelson Roberto Furquim © 2018 Nova Science Publishers, Inc.

Chapter 8

PACKAGING OF RED MEAT

José Boaventura M. Rodrigues[1]
and Claire Isabel G. L. Sarantópoulos[2,]*

[1]DuPont Performance Materials, Barueri, Brazil
[2]Centro de Tecnologia de Embalagem,
Instituto de Tecnologia de Alimentos, Campinas, Brazil

ABSTRACT

Beef packaging plays a key role in maintaining product quality and safety throughout the supply chain, to promote the product and convey relevant information to consumers.

Various aspects related to the packing system must be considered while specifying beef packaging, given their straight correlation with operating costs, productivity, quality, shelf life and compliance with trade regulations. Increasing consumer awareness and environmental demands challenge beef industry stakeholders to enhance food security, reduce losses, waste and the environmental footprint. Packaging is one critical element affecting these requirements.

---

[*] Corresponding Author Email: claire@ital.sp.gov.br.

Besides providing a reliable barrier between the product and the outer environment, packaging maintains the required atmosphere to preserve the product and allow biochemical processes of fresh meat to occur unrestricted, resulting in desirable organoleptic characteristics to meet consumers´ needs such as attractive color and tenderness. Each packaging category has specific advantages and disadvantages that must be considered when selecting a system to provide an adequate balance of protection (shelf life), shelf appearance, cost and environmental implications.

This chapter describes the main features and expected general performance of currently available packing systems and materials and it also briefly describes promising novel packing technologies. Major market trends are discussed to support designing and managing beef packing system processes.

Keywords: vacuum pack, modified atmosphere packaging, vacuum skin pack, beef preservation

INTRODUCTION

If unprotected, beef quality will rapidly decay. Growing urbanization and expansion of global trade demand measures to preserve beef products. In addition to extending the shelf life, brand owners and retailers must find effective ways to present their products in a neat and convenient way, giving an appearance that shows it is safe and nutritious.

In the early 1900s, fresh bare meat carcasses were sent to butcher stores to be processed into cuts and packed under consumers' supervision. The advent of 'boxed beef' made distribution and storage easier and retail stores were supplied with meat already trimmed into selected cuts. By the mid-20th century, the introduction of individual vacuum-packed cuts allowed meat processors to control product aging, therefore meat would be consumed after proper maturing was achieved. The next step of this evolution was to develop case-ready meat, packed at central processing plants and delivered to retailers ready for sale, with high level safety, high quality standard and branded. By 2010, about 30% of beef and 70% of ground beef were sold in case-ready systems in the United States (NAMI, 2015).

Color is the main feature referred to by consumers to judge beef quality. The chemical state of myoglobin, the morphology of the muscle structure and the ability of the muscle to absorb or scatter incident light determines the color of the red meat. The presence of liquid exudate, also referred to as "purge" or "drip," also affects the appearance of the product. Up to 2% liquid exudate found in packaged beef is considered acceptable, while 4% is excessive and incurs substantial economic losses (Robertson, 2013). Further losses occur when meat is cut because exposed surfaces increase, exacerbating liquid purge, proteolytic and oxidative reactions.

The aim of packaging for fresh muscle foods is to inhibit specific chemical reactions, microbial growth and physical changes, consequently delaying undesirable changes to the appearance, flavor, odor and texture, but not interfering with enzymatic activities that are beneficial (Scetar et al., 2010) to enhance tenderness, juiciness and flavor.

In this regard, packaging helps to protect meat from discoloration, development of off-flavor and nutrient loss (Jin et al., 2013) extending the shelf life. Packaging is also the media used to promote the brand and convey information such as: meat processor, nutrition facts, manufacturing and expiring dates, origin and other distinguishing features that might favor the selection of the product.

Notwithstanding its importance, packaging alone fails to preserve red meat given that endogenous factors also influence deterioration, especially fresh products. Therefore, packaging must be regarded as one element of a system that also concerns: characteristics of livestock (breed, age, nutrition); the sanitary state of pre-packed products; temperature; gas composition; and light exposure.

Although enzymatic deterioration (e.g., proteolysis, lipolysis) and oxidation can occur in the absence of microorganisms, microbial growth is by far the most important factor affecting the quality of fresh meat (Zhou et al., 2010). If adequate packaging and/or environmental conditions are not in place, the color, odor, texture and flavor of meat will deteriorate.

FACTORS AFFECTING RED MEAT QUALITY

Quality loss of red meat is associated with pigment oxidation, microbial growth, rancidity, superficial dehydration and drip loss. These deterioration processes are influenced by temperature and intrinsic and extrinsic factors such as gas atmosphere, light and relative humidity. As pointed out earlier, packaging works as part of a system to protect flesh products, as well as temperature and good manufacturing practices.

Fresh red meat is one of the most perishable foods and refrigerating temperatures are always used to delay quality losses. Therefore, temperature control is one key element of concern for the red meat industry, because it controls microbial growth, chemical and enzymatic reactions. Low temperatures are used for storing carcasses after slaughtering, primal and subprimal cuts, case-ready packs and during transportation, storage and retail operations. Therefore, beef must be maintained chilled at temperatures within -1.5°C and 4°C or frozen at temperatures below -15°C.

Microbial spoilage leads to undesired changes in odor (volatile metabolites), color and appearance (slime formation). These organoleptic changes may vary according to the type and number of microorganisms contaminating the meat and the storage conditions. Many of these microorganisms potentially contribute to meat spoilage under appropriate conditions. This makes the microbial ecology of spoiling raw meat very complex, and thus preserving it difficult. The microbial growth can be managed using controlled temperature and gas atmosphere around the product.

An aerobic environment favors the development of microorganisms with higher potential for spoilage such as Pseudomonas. It also induces oxidation of pigments and fats, resulting in color and flavor changes. On the other hand, anaerobic environment as in vacuum packages, microorganisms of lower deterioration potential prevail, such as lactic acid bacteria. However, in the absence of oxygen (O_2), beef pigment myoglobin acquires a purple color, which may limit its acceptance in the retail market.

Alternatively, microbial growth can also be controlled by intermediate to high levels of carbon dioxide (CO_2), which has an antimicrobial effect.

Vacuum packaging and modified atmosphere packaging - MAP, modify the atmosphere to which the microorganisms are exposed, altering the growth rate and also the composition of the meat microflora. By doing this, the time and the type of deterioration can be controlled, and consequently the shelf life can be extended.

Color is an important marketing attribute of red meat, both in the chilled and frozen form. Consumers associate color with meat freshness. The meat color is determined by the concentration and chemical form of the meat pigment myoglobin, the morphology of the muscle structure and its capacity to scatter or absorb light. Myoglobin has a heme group with a central iron atom that has six bonding points or coordination links. One of these linkages is free to bond oxygen, water or other substances such as carbon monoxide (CO), which will determine the color of the pigment. Other factors that determine the pigment color of meat are: the oxidation state of the iron atom (reduced ferrous form or oxidized ferric form) and the state of globin, the globular protein of the pigment (native in raw meat or denatured in cooked meat).

The color of fresh meat depends on the relative amounts of deoxymyoglobin (Mb), oxymyoglobin (O_2Mb) and metmyoglobin (MetMb). Mb is purple and predominates in the absence of O_2, as in vacuum packaging and in high carbon dioxide (CO_2) MAP. O_2Mb is a bright red color and is present when the meat surface is exposed to intermediate to high O_2 atmospheres, such as air or high O_2 MAP. This is the "bloom" of fresh red meat. MetMb is the oxidized form of myoglobin. It is brown and occurs at low O_2 concentrations or when red meat is exposed to air for some time. The interrelationships between the three forms of the myoglobin is reversible and dynamic. Although MetMb is stable and cannot take up O_2, it is slowly converted to Mb in *post rigor* meat. Carbon monoxide (CO) combines with myoglobin to form carboxymyoglobin (COMb), the bright cherry red color, which is similar to the O_2Mb color.

The interrelationships between the three forms of the myoglobin are used in different approaches of MAP for fresh red meat. The purple Mb is more stable than the bright red O_2Mb. The bright cherry red COMb is much more stable in terms of oxidation than O_2Mb, even at very low levels. Therefore, the color stability of vacuum packaging (3 to 4 months at 0°C) is higher than the high O_2 MAP (10 to 15 days). The shelf life of MAP with CO depends on the microbiological stability of the product and not the color.

Regarding the control of the atmosphere around chill-stored red meat, vacuum packaging depletes oxygen around the product minimizing microbial and chemical deterioration processes and consequently extends the storage life of chilled meat. Modified atmospheres packaging (MAP) also extends the storage life by replacing the air surrounding the product by a mixture of gases such as: carbon dioxide to restrain bacterial growth and high oxygen to maintain the red color or, in some cases, carbon monoxide to improve color stability (Luzardo et al., 2016).

Although the shelf life of frozen beef can be extended beyond 12 months, oxidative reactions and physical changes still continue affecting quality with time. The most obvious deteriorative effects related to freezing are: surface dehydration; fat and pigment oxidation; and mechanical damage occurring throughout product handling and transportation. Surface dehydration (caused by water sublimation), popularly known as 'freezer burn', leads to a concentration of solute, resulting in undesired color and an increased rate of oxidation of pigments, fat and vitamins. The packaging used must have a high moisture barrier and small headspace to prevent dehydration and withstand mechanical damage such as punctures and tear. To minimize oxidative reactions, air must be exhausted from the package, and therefore the packaging must be flexible enough to tightly wrap the product and maintain hermetic seal at low temperatures.

Table 1 summarizes the effects of critical parameters on chilled or frozen beef quality and its relationship with packaging.

Table 1. Critical parameters on chilled or frozen beef quality

Critical parameters	Effect	Associated problems	Relationship with packaging
Muscle and fat quality (pre-packed product quality)	Microbial counts, color and flavor stability and texture	Off-odor and gas production (microbial deterioration), discoloration (pigment oxidation), off-flavor and loss of meat tenderness with impact on quality and shelf life	None
Temperature	Enzymatic and chemical reaction speed, microbial activity and physical changes	Off-odor, discoloration, dripping with impact on quality and shelf life	Packaging must allow fast cooling and resist mechanical damage at low temperatures
Surrounding gas composition	Microbial growth and chemical reactions	Discoloration and off-odor with impact on quality and shelf life	Packaging maintains the designed gas atmosphere either allowing or restraining gases to permeate inside or preventing gas to escape from headspace
Light	Color (chemical state of myoglobin)	Discoloration	Packaging can be a barrier to UV light (anti-UV additives or printing)

SELECTING THE PACKAGING SYSTEM

Various packing systems are available for use in the meat industry, distribution centers or retail stores but none are capable of thoroughly addressing the attributes required to preserve, transport and present the product all at once. The most effective packing system has to be designed or selected based on the optimal net benefit, balancing the bonuses of each system is expected to deliver in light of business strategies, customer needs and trade regulations. Considering all this, there will be cases where more than one system is needed to fulfill the requirements of different stages throughout the supply chain.

Quite a few key factors must be considered while selecting a packing system for meat products. Attributes related to shelf life, commercialization temperature, packing productivity and costs are of utmost importance for manufacturing. Packaging design (volume, shape, weight and appearance), labeling, pricing, regulatory information, disposal of post-consumer packaging and lifecycle footprint are required package selling specifications.

Figure 1 depicts the value chain of beef, showing different types of packaging used for wholesale distribution and retail. Wholesale packaging preserves red meat since the product is packed until it is opened and the product is transferred to the retail unit. Retail packaging is considered the system used to display products for selling them to consumers, regardless of whether it is done by meat processors, distribution centers or in retail stores. Wholesale packaging focuses primarily on extending the shelf life, meaning preserving the quality of the flesh product. Retail packs are responsible for preserving quality, but also address customer requirements related to appearance, information and convenience.

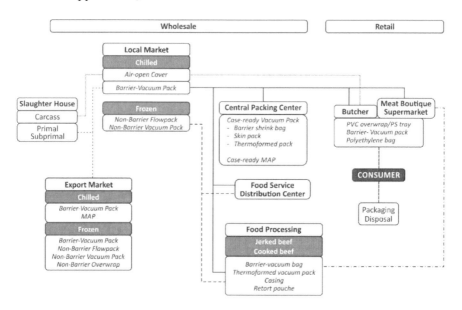

Figure 1. Value chain of beef.

PACKAGING CATEGORIES

Plastic packages for flesh products are divided into many different categories: air open covers; film overwraps (shrunk or not); vacuum packs; modified atmosphere packaging; plastic bags; casings; and retort containers or pouches.

Air Open Covers

Although this packaging category is becoming obsolete, it is still used in developing countries for beef wholesale distribution to butcher stores located nearby slaughterhouses. The system consists of a polyethylene protective liner covering the carcass or primal cuts to serve as a barrier to dust and cross contamination of beef while transporting or handling under refrigeration. Open covers do not provide protection against purge and moisture loss and have very limited ability to extend the shelf life of the product. The advantage of this system is its low cost since no fancy materials or technologies to manufacture the film are required and the lining operation is made manually with little capital expenditure.

Retail Overwraps and Bags

Plastic overwrap films are used for making flesh products available for consumers.

Poly vinyl chloride (PVC) overwrap films are usually used by retail shops that receive meat in open air covers. The meat is cut and placed over a rigid plastic tray and then wrapped with a thin transparent PVC stretch film for display on chilled shelves. Plastic trays consist of a thermoformed polystyrene (PS), expanded PS foam (EPS), PVC or polyethylene terephthalate (PET) recipient that holds the product, as well as any drip coming from meat.

This system has the advantage of using small size, low cost and easy to operate equipment. It also provides a moisture barrier to minimize weight loss and is a reliable safeguard to avoid contamination when the product is handled or transported from the shop to home. Since the PVC film is held in close contact with the surface of red meat, it allows oxygen to permeate inwards to the bright red color of oxymyoglobin bloom, giving the product an appealing appearance. Such a packaging system does not prevent purge, therefore a liquid absorber is usually placed between the meat and the tray, to prevent liquid from flowing through the tray, eliminating the undesirable appearance of drip.

PVC overwraps preserve steak for 5 to 7 days at very low chilled temperatures but shorter periods for ground or minced beef before surface browning happens due to oxidation of myoglobin into metmyoglobin even if low bacterial count is found. At higher temperatures shelf life is limited to 2 to 3 days. Spoilage bacteria growth, such as *Pseudomonas*, is favored in PVC wrapped beef as compared to environments deprived of oxygen. This microbial growth is accompanied by off-odor, slime and greenish spots on the meat surface. PVC film is also prone to punctures and tearing, leading to frequent leaky packages (Gazalli et al., 2013).

Gas permeable polyethylene bags are used by butcher shops to pack meat as required by the consumer. They are intended for short-term storage at home, providing a minimum degree of dust, liquid and moisture barrier to mitigate weight loss and maintain freshness and tenderness of meat stored in domestic fridge temperatures.

High Permeability Flowpacking

High oxygen permeability stretch-shrink polyolefin films are used for packing fresh beef primal cuts using flowpack technology. These cases may or may not use rigid trays to accommodate the cut. This system is similar to those used for poultry, however the packing film must enable very high gas permeation to allow beef surface blooming and reduce discoloration risks.

Masterpacks

Retail high permeable overwraps and flowpacks are consolidated in masterpacks for distribution purposes where vacuum or modified atmosphere with high CO_2 concentration is applied to extend product shelf life. In either cases, beef develops purple color. After arriving at the retail store, the masterpack is opened and the primary packaging is exposed to atmospheric oxygen, consequently developing a bright red color so as to be attractive to consumers.

Vacuum Packaging

Depriving red beef from oxygen extends its quality by reducing the activity of spoilage flora that is dominated by species that grow under aerobic conditions. At chill temperatures and anaerobic conditions, lactobacillus species outgrow spoilage aerobic competing species. Lactobacillus has a lower potential for deterioration as compared to aerobic microorganisms, significantly increasing the meat shelf life. For chilled beef maintained under vacuum, the optimum temperature range is $1.5 \pm 0.5°C$ (Robertson, 2013).

For any packing system that preserves beef under vacuum, it is imperative to eliminate as much air as possible from the package as oxygen levels, even as low as 0.15% to 0.20%, predispose the product to browning (Scetar et al., 2010). Therefore, making sure there is close contact between the meat surface and the packaging material and preventing air from re-entering the packaging is of utmost importance. The packaging structure must be designed to be flexible (to totally wrap the product), puncture resistant, assure hermetic sealing and be impervious to gases. The ability of certain films to shrink and wrap around the product and minimize dripping *in* the pack makes them a popular choice for beef packaging. Puncture resistance becomes critical for packing bone-in and frozen beef due to the possibility of hardened tips of bones and frozen flesh to perforate the film during transportation and storage.

After evacuation, any remaining oxygen inside the packaging, including O_2 dissolved in the product, will be consumed by enzymatic reactions within the muscle tissue and microbial activities replacing it with CO_2 that has a desired effect to inhibit microbiological growth (Toldra, 2010).

Vacuum packaging, followed by refrigerated storage, is the most effective method currently used to extend shelf life of uncooked meats (Jin et al., 2013). Chilled beef with pH below 5.8 can be kept edible for as long as 14 weeks at 0°C (Robertson, 2013) and even longer periods for frozen beef. This system also allows fresh beef tenderness to evolve naturally by enzymatic reactions and minimizes the loss of weight due to the evaporation of surface moisture.

The biggest disadvantage of vacuum packed red meat is that as oxygen is removed from packaging, oxymyoglobin can no longer prevail on the muscle tissue. In this condition, most of the myoglobin pigment will be in the state of deoxymioglobin giving beef a purple color. Since color is one key factor defining beef acceptance by consumers, vacuum packaging is more commonly used for wholesale trades of primal cuts.

Regardless of the appearance disadvantage, vacuum packaging has also become a popular system for retail packaging since its impact on the increase in shelf life is significant. For that purpose, subprimal cuts are packed taking advantage of existing equipment at beef processing plants. This is also regarded as an option to reduce risks of contamination throughout intermediate processing.

There are numerous versions of vacuum packaging systems. The most common ones are: barrier-non-shrink bags, barrier-shrink bags, barrier thermoformed packs, barrier skin packs and non-barrier skin packs.

Barrier-Non-Shrink Bags

A subprimal beef cut is placed inside a preformed bag with three welded edges and one open edge (mouth). The meat inside the bag is placed in a vacuum chamber so that the open edge of the bag stands over the sealing bar. The chamber is closed and vacuum applied. When the preset vacuum pressure is achieved the sealing bars close and heat is applied sealing the bag. Afterwards the internal pressure inside the chamber is normalized with the ambient, the chamber is opened leaving the packages tightly formed around the meat surface.

This system is used for jerked beef and cuts with flat surfaces, with squared shape and straight corners. The use of softer materials such as ethylene-vinyl acetate copolymer (EVA) or ionomer as the inner layer of a multilayer film is recommended to pack round cuts and meat that releases liquid purge. These polar resins with low melting point inside the package stick together forming what is popularly known as secondary sealing. Secondary sealing prevents drip to flow or the cut to move freely inside the packaging, giving it a better appearance.

Vacuuming is used adopting single or dual vacuum chamber equipment for manual or semi-automated operations or multichambered rotating automatic equipment.

Barrier-Shrink Bags

Barrier-shrink bags is probably the most popular system used to pack fresh beef and is still the most cost-effective packaging strategy adopted for packing meat (Toldra, 2010) provided it fits for packing regular or uneven cuts of bone-in or boneless cuts in fresh, cured or frozen states.

The packing process uses manual, semi-automated or automated machines, similar to those described for non-shrink bags, with the exception that after beef is vacuum packed the package goes through a shrinking process in a hot water bath or heating tunnel at approximately 90°C for no longer than one minute.

The shrink ability of packaging is given by residual tension added to the film using a technology called double-bubble coextrusion. After the hot bubble is formed, it is stretched twice along the machine and the transverse directions and quenched, crystalizing tensions to the solidified film.

Tensions will be relieved while heated after the meat is vacuum packed making the film shrink and enclosing the meat tightly.

Usually it is labeled placing an internal printed plastic tag inside the packaging. Printing or sticking labels directly to the packaging is challenging given the figure will distort throughout the shrinking process.

Thermoformed Packaging

This system consists of a continuous sequence of operations performed by a piece of equipment with or without human assistance. A plastic web is preheated, vacuum formed into trays, filled with the product and moved to a vacuum chamber where another plastic film is placed on top and the package is heat-sealed under vacuum. Top film (lid) and bottom film (thermoformed flexible tray) have distinct performance requirements, although both must present high gas barrier properties.

The advantages of this system are increased productivity as compared to the above manual bagging operations and the neat appearance of the package that allows the lidding film to be printed or labelled for branding the product. This system is mostly used for premium fresh, frozen, cured or dried cuts that can be made available in standard sizes as the thermoforming operation allows for narrow variability of cut dimensions and weight.

Vacuum-Skin Packs (VSP)

This is also an automated or semi-automated system and comprises a rigid plastic tray supplied preformed or thermoformed inline. The meat cut is placed in the middle of the trays leaving a fair distance from the edges. A preheated softened plastic sheet is put on the meat in the tray under vacuum, taking the shape of the product. When the vacuum in the chamber is undone and the packaging returns to atmospheric pressure, the top film

Packaging of Red Meat 207

collapses onto the product, completely wrapping the surface of the flesh as if it were a second skin.

Vacuum skin packages are mainly used for small portions, fresh, processed or frozen red meat. They give a neat appearance to the product, and printed branding is allowed stuck onto the back of the rigid tray. Since the product is maintained tightly wrapped and surrounded by the skin film, there is no or very little dripping and the package can be safely displayed vertically on the shelves with little jeopardy to the display.

Depending on the depth of the rigid tray, another way to close the package is to heat seal another transparent cover over the edges of the tray. For this case, an oxygen rich gas mixture fills the space between the surfaces of a gas permeating skin film and the top gas barrier lid. Since oxygen is allowed to permeate through the skin film, oxymyoglobin blooms a bright red color on the surface of beef displayed on shelves.

The skin film is required to be: transparent and glossy to enhance presentation; soft and tough to mold around the product; and to be puncture resistant in case of bone-in cuts. The film surface presented to the tray and meat must be compatible with the tray's surface, resulting in a reliable air tight closure.

Vacuum skin packaging is intended for retailing standardized premium cuts. The operation can be performed at beef processing plants or distribution centers, but less likely at retail stores given the investment required is significant. The fresh product shelf life can be 15 – 22 days (Toldra, 2010) depending on the beef cut, temperature and packaging feature (barrier or non-barrier skin film, with or without modified gas atmosphere).

For frozen beef, a permeable skin packaging system can be used given that the below freezing point temperature inhibits microbial deterioration. Low temperature conditions also delay fat and pigment oxidation, even with the presence of atmospheric oxygen, and the packaging prevents the freezer burn.

Shrink Flowpacks

Flowpacks for vacuum applications are typically a horizontal, fill and vacuum seal process that allows a high degree of automation, higher packing speeds and minimizes leakage in the sealing area and weight of packaging required, as compared to the previous described vacuum systems, giving a neat presentation to the product.

Beef cuts are fed into a conveyor belt. An in-line product presence detector releases a plastic film that surrounds the product while a heat and pressure gauge seams the film horizontally. As the surrounded cut moves along with the belt, the back part of the film is cut and heat-sealed, leaving only the front end open. The package enters a vacuum chamber where it stops for a while so that the air can be taken out and a last seal is made at the front end of the film. After pressure is normalized, the film wraps the product, then it finally goes through a heating tunnel to shrink the film even more tightly.

This system may be regarded as an automated alternative to shrink-bags with similar performance regarding protection, appearance and shelf life. The trade-off is that each machine has limitations to accommodate wide variations of cut widths. One processing line will operate as many flowpacking lines as necessary to adjust to different beef cuts.

As mentioned for the previous cases, packaging will rely on good moisture barrier films. High gas barrier films may be preferred to pack fresh chilled beef, while medium or low gas barrier films may be used to pack frozen beef, depending on the desired shelf life. Film structures are similar to those described earlier for shrink bags.

Modified Atmosphere Packaging – MAP

Initial attempts to use gas, such as carbon dioxide for red meat, was aimed exclusively to extend the shelf life of beef carcasses for long distance trade. More recently, modified atmosphere packaging (MAP) has also been used as an alternative to vacuum packaging to preserve meat

Packaging of Red Meat

while maintaining a bloomed red color and providing a convenient case-ready packaging.

In this regard, MAP is becoming a popular choice for consumers who can afford products branded for its convenient serving sizes to be kept for a few days in a domestic refrigerator with the appearance of a fresh product.

Studies found in the literature still debate what is the most appropriate gas composition and conditions in which MAP is advantageous compared to traditional retail packages such as high permeable PVC overwraps and vacuum packaging.

Modern MAP for fresh beef is defined as a rigid tray and a hermetically sealed flexible lid. The package must present moisture and gas barrier to retain the internal atmosphere over the timespan from packing to consumption. The tray must be high enough to ensure the product will fully fit inside the internal volume and the top surface will not touch the lid so that the gas blend will be fully available to the meat surface.

Alternatively, trays with steak or minced beef covered with breathable film are placed inside a gas impermeable masterpack and air is replaced with modified atmosphere gas flushed.

The additional 7 to 10 days of red color stability justifies adopting retail MAP carried out at central packing facilities in large volumes compared to meat overwrapped in PVC films at local butcher stores (Gazalli et al., 2013). A centralized packing operation is required for MAP due to the safety precautions required to handle gases in confined spaces.

Given the need to maintain gas composition within certain limits, the packaging structures used on MAP must prevent gas exchange in and out the packaging. On the other hand, it is impossible to guarantee steady gas composition inside the packaging given that the internal atmosphere will be modified by natural metabolism of *post mortem* meat during storage.

Therefore, initial gas composition and the relative volumes of gas and meat are important variables to be considered in determining the progress of the changes in gas concentrations during storage.

The most popular MAP system uses a blend of O_2 and CO_2 to preserve fresh beef. Oxygen concentration higher than 21% (normal atmospheric concentration) provides the product with a bright red pigmentation while CO_2 has a beneficial bacteriostatic effect, limiting microbial deterioration.

Nitrogen (N_2) can be also used as an inert gas blended to CO_2 and O_2 to equalize internal packaging with external ambient pressure. Retail ready sub primal cuts, steak or ground beef are packed using this system.

Using O_2 volumetric concentrations as high as 75% to 80% and CO_2 20% to 15% at chill temperatures (below 4°C) preserve fresh red meat up to 10 to 12 days, maintaining bloomed red color associated with the presence of oxymyoglobin on the surface of beef. However, there have been reports of off-odors and rancidity in meats stored at high O_2 concentration. Studies described beef steaks in high O_2 MAP showed significantly lower tenderness and juiciness, and higher off-flavor scores compared to steaks in vacuum packaging, suggest that the high O_2 MAP system may induce lipid and myoglobin oxidation and cross-linking and/or aggregation of myosin by protein oxidation (Robertson, 2013).

The gaseous volume of the pack should be around twice the volume of the packed meat to avoid excessive variation of atmosphere composition. The red meat should not touch the lid ensuring the surface (that the consumer sees) will be fully oxygenated.

In MAP with low oxygen concentration, most air is evacuated from package and replaced by pure CO_2 or CO_2/N_2 blend where CO_2 minimizes microbial growth and N_2 (when applied) is an inert gas added to regulate internal pressure and maintain the shape of the package. Residual O_2 inside the package will eventually be absorbed or used up by natural meat metabolism. Myoglobin will be present in the deoxymyoglobin state, rendering a purple color to the beef and making this system more suitable for wholesale transportation of primal cuts, rather than the case ready for retail sales. Considering this, the effect of low O_2 MAP on red meat preservation may be regarded as similar to vacuum packaging. Some authors point out that low O_2 MAP can reduce liquid purge from meat since the product is not subjected to mechanical strain as in vacuum packs (Robertson, 2013).

Although controversial, adding CO at 0.4% to the gas mixture is allowed in countries such as the United States. The presence of CO

Packaging of Red Meat

improves color allowing the adoption of low O_2 MAP for retail ready packaging. Using CO can also blend higher CO_2 concentrations to the gas mixture to control microbial deterioration.

The setback is that CO may conceal spoilage at the selling point if the system fails to maintain the temperature below 4°C. In this case, the spoilage will only be noticed after the package is opened and the off-odor reveals the deterioration.

Ultra-low O_2 MAP ensures that only tiny amounts of O_2 in the modified atmosphere are kept inside the packaging filled with CO_2 and/or N_2 to prolong shelf-life of beef in master (bulk) packs for as long as 4 months when kept at -1°C. In this case, the packaging material is typically a flexible bag presenting very low water vapor and gas permeability that is filled with the product using equipment that removes the air from the packaging by applying vacuum. The packaging is gently squeezed to eliminate free space within the product followed by a flush of gas mixture that is free or has traces of oxygen. After the designed gas volume is filled, the bag is tightly sealed and placed into secondary rigid cardboard box to protect the product while it is being transported.

To avoid browning of beef due to oxidation of myoglobin into metmyoglobin, oxygen levels inside the packaging must be less than 0.15% (Scetar et al., 2010). Ultra-low oxygen concentrations can be achieved by using oxygen scavengers in the packaging.

A minimum of 20% CO_2 is generally required to control the bacterial growth. However, concentrations of 100% CO_2 may be used in bulk packs of meat (Mullan et al. 2003). CO_2 is highly soluble in muscle and fat tissues. Therefore, if CO_2 is used in concentrations above 70%, an excess gas volume of at least similar to the volume of packed meat is required to avoid the packaging from collapsing and maintaining muscle tissue in a saturated state. An insoluble gas, such as nitrogen, may be added to the gas mix. The disadvantages of this requirement are: an increased bulk volume, leading to extra transportation and storage costs; and off-flavors of meat if degassing is not allowed for at least 30 minutes in open air. CO_2 is also thought to affect the quality of meat by lowering its pH resulting in protein denaturation and reducing the ability of muscle to retain water.

Retort Pouches and Metal Cans

Retort packaging is used for very long-term shelf-stable storage of cooked beef and ready meals. The shelf stability may last for years. This system has been developed primarily as a means to provide secure and nutritious rations for the military, but has also been used as a convenient way to store beef for general consumers, especially where there are no cold chains or they are economically unviable.

In this system, beef is kept in sterile conditions – eliminating any bacterial or enzymatic activity by treating the product already packed with temperatures ranging from 110°C to 160°C. Throughout the heat treatment, beef is partially or completely cooked promoting chemical and physical stability at room temperatures.

Since the product is treated inside the package, the packing materials must resist cooking temperatures without allowing migration of chemical compounds to the product above legal limits. Metal cans or flexible pouches, also known as 'retort pouches,' may be used for this purpose.

Tinplate and chromium treated steel cans are widely used for shelf-stable beef since they allow hermetic closure and provide a barrier to oxygen and light avoiding microorganisms from entering the package and granting commercial sterility of the content after thermal processing. Metal cans can also stand cooking temperatures, having high mechanical resistance and can be easily discriminated from other post-consumer packaging materials to be recycled. Lacquer is applied to the internal surface of the can to prevent direct contact between the product and the metal to avoid metal corrosion.

Closing metal cans consists of a lateral seam with solder or electrical welding. The cans may be supplied with a molded bottom (one piece molded can) or with a crumpled bottom fix. A metal cap is fixed after the product is fed into the container and before it is cooked.

Retort pouches are multilayer films made of plastic materials (in some case coated with silicon oxide – SiOx or aluminum oxide - AlOx) or a combination of plastic and aluminum foil that provides gas, moisture and light barrier and hermetic sealing to maintain the product in sterile conditions. This flexible packaging system has become popular for packing wet pet food.

Given plastic seals will not resist high pressure at high temperatures, heat treatment is applied inside pressurized autoclaves after pouches are placed in compartmentalized shelves to restrict the package reducing risks of damaging the seal. The shelves are perforated to grant homogeneous temperature inside the equipment. The autoclave pressure must be set-up and controlled throughout the whole heating processes to avoid the relative pressure of pouches to cause damage to the seal.

PACKAGING MATERIALS

Packaging fresh red meat works as a barrier to minimize weight loss and maintain an internal environment that does not favor harmful microbiota growth, preserve organoleptic properties of the product.

Contemporary packaging systems most often rely on distinct materials in five or more layers to form a flexible film or rigid tray. In general, the external layer provides a glossy finish and abrasion resistance to enhance and maintain the package's appearance. The inner layer heat-melts and welds on itself or to other materials to assure a hermetic seal. The layers in-between provide mechanical resistance, toughness and a barrier to minimize gas and water vapor permeation.

Polymers Used in Red Meat Packaging

Poly Vinyl Chloride (PVC)
Films made with PVC have been used to pack meat cuts since the 1940s. The stretch ability and high gas permeability of plasticized PVC films allow close contact with the product and the oxymyoglobin to prevail on the surface of fresh and frozen beef, giving a blooming red color to the product. The film also minimizes moisture loss, provides a barrier against exogenous microbes and neat product presentation.

Polyethylene (PE), Ethylene Copolymers and Ionomers

Polyethylene and ethylene copolymers are usually used as outer, core and (inner) sealant layers to provide the packaging with toughness, puncture and seal strength and as a barrier to minimize weight loss. These polymers, although permeable to gases, have the highest water vapor barrier among materials used to manufacture red meat packaging.

Ionomers, such as DuPont™ Surlyn®, are widely used as sealant layers given their low and wide range heat seal initiation temperature and their ability to maintain a reliable seal while hot – also called hot-tack sealing (Scetar et al., 2010). While used as a core layer, ionomers contribute to improving puncture resistance and shrink ability on films made from double bubble coextrusion (Rodrigues et al., 2017b).

Polyamide (PA)

Polyamides are good barriers to gases and organic vapors, present high mechanical resistance, good thermal resistance, good barrier to fat, good flexibility at low temperatures and can be thermoformed. They are sensitive to moisture so they lose part of the mechanical and barrier properties with humidification. Polyamide homopolymer (PA 6) and copolymers (PA 6/66) are the primary types used in flexible packaging both with good oxygen permeation performance (Morris, 2017). They can be used pure or blended. Blending PA 6 or PA 6/66 with amorphous PA (amPA) can improve transparency, processing and gas barrier at higher moisture content (Rodrigues et al., 2017a). More recently, terpolymers are being used for high performance applications. They are distinguished by high transparency, greater flexibility and puncture resistance, better performance in shrinkage and a lower melting point. They are recommended for shrink bags for meat with bones, deep thermoformed packages and artificial casings.

Polyvinylidene Dichloride (PVDC)

PVDC presents excellent gas and water vapor barriers and the halogen groups attached to the polymer interact more strongly with each other than

Packaging of Red Meat

with water, making gas permeability insensitive to water or environmental moisture (Morris, 2017).

Typical structures for fresh and cured beef places PVDC as a middle layer in combination with side layers of EVA (ethylene-vinyl acetate) and PE to provide toughness, seal ability and high moisture and oxygen barrier properties.

Ethylene Vinyl Alcohol (EVOH)

EVOH comes in a variety of proportions between the ethylene and vinyl alcohol monomers ranging from 27% to 48 mol% of ethylene. The lower the percentage of ethylene, the higher the barrier (i.e., the lower the oxygen permeation). Although EVOH presents very low gas permeability, it is highly sensitive to moisture, given that water molecules bind to vinyl alcohol groups, plasticizing the polymer matrix and creating voids through where oxygen molecules can permeate (Morris, 2017). Therefore, oxygen transmission rates reported in the literature must be used with caution for meat packaging applications.

To mitigate moisture sensitiveness, EVOH is usually encapsulated within layers of moisture barrier materials such as PE, PA and PET (polyethylene terephthalate). Gas barrier films made with EVOH are usually made of 7 up to 11 layers of materials that provide the required moisture and gas barriers, seal resistance and toughness. These films are used for vacuum packs and MAP.

Packaging Structures

Conversion processes of plastic packaging such as lamination, coextrusion and coatings allow the combination of several materials in multiple layers. These technologies enable the optimization of packaging properties at a minimal cost. Many flexible and rigid plastic structures can be used for different meat packaging systems.

High permeable (or low barrier) films are: PVC monolayer stretch films with various level of plasticizer used as plastic overwraps; and multilayer coextruded polyolefin based films used as plastic bags, flow packs and skin packaging.

Medium and high barrier films are commonly used for: barrier shrink and non-shrink bags; thermoformed packaging; MAP and master packs.

Coextruded structures may incorporate external printing or labeling for product identification. Normally, thermoformed packaging lids show reverse printing. In this case, the coextruded films are laminated to an external film with reverse printing. This type of material is called hybrid films, because it combines two conversion processes: lamination and coextrusion.

Table 2. Examples of barrier-non-shrink bag structures

3-layer coextruded structures	Gas barrier level	Overall gauge (μm)
PA/tie/PE	Medium	90 - 120
5-layer coextruded structures	**Gas barrier level**	**Overall gauge (μm)**
PE/tie/PA/tie/PE	Medium	70 - 110
PE/tie/PA/tie/EVA	Medium	70 - 110
PE/tie/EVOH/tie/PE	High	70 - 110
7-layer coextruded structures	**Gas barrier level**	**Overall gauge (μm)**
PE/tie/PA/tie/PA/tie/PE	Medium	70 - 110
PE/tie/PA/EVOH/PA/tie/PE	High	70 - 110
9-layer coextruded structures	**Gas barrier level**	**Overall gauge (μm)**
PE/PE/tie/PA/PA/PA/tie/PE/PE	Medium	55 - 110
PE/PE/tie/PA/EVOH/PA/tie/PE/PE	High	55 - 110
Hybrid structures (laminated+coextruded)	**Gas barrier level**	**Overall gauge (μm)**
PET/adh/PE/tie/EVOH/tie/PE	High	70 - 110
PET-PVDC/adh/PE	High	70 - 110
5-layer coextruded structures [1]	**Layer thickness (%)**	**Overall gauge (μm)**
PA/EVOH/tie/LLDPE/mLLDPE[2]	15/10/15/30/30	50 - 120
PA/EVOH/tie/LLDPE/ION [3]	15/10/15/30/30	50 - 120

[1] Adapted from MORRIS, 2017.

[2] mLLDPE – linear low-density polyethylene made with metallocene catalyst.

[3] ION - used to provide secondary seal ability.

Tables 2 to 6 present various commercial structures or multilayer barrier films. In these tables, 'tie' means a coextrusion polymeric adhesive,

Packaging of Red Meat

such as maleic anhydride modified polyolefins and 'adh' are lamination adhesives that can be solvent, waterborne or solventless that are applied to the surface of a film to adhere to another film.

Table 3. Examples of barrier-shrink bag structures

5, 7 and 9-layer coextruded structures [1]	Layer thickness (%)	Overall gauge (μm)
EVA/tie/PVDC/tie/mLLDPE	40/5/10/5/40	50 - 120
EVA/tie/PVDC/tie/EVA/EVA/mLLDPE	30/5/10/5/15/15/20	50 - 100
PET/tie/ION/tie/PA/EVOH32/PA/tie/mLLDPE	10/5/30/5/5/5/5/30	50 - 100
3, 5, 7 and 9-layer coextruded structures	**Gas barrier level**	**Overall gauge (μm)**
EVA/PVDC/EVA	High	55 – 100
EVA/EVA/PVDC/EVA/PE	High	50 - 100
PE/EVA/PVDC/EVA/PE	High	50 - 100
PE/EVA/EVA/PVDC/EVA/EVA/PE	High	50 - 100
PE/EVA/EVA/PVDC/EVA/EVA/ION	High	50 - 100
PE/tie/EVOH/tie/PE	High	50 - 100
PE/tie/PA/tie/PE	Medium	50 - 100
PET/tie/EVA/tie/PA/EVOH/PA/tie/PE	High	50 - 100
PET/tie/ION/tie/PA/EVOH32/PA/tie/ION	High	50 - 100

[1] Adapted from MORRIS, 2017.

Table 4. Examples of barrier thermoformed structures

7 and 9-layer coextruded structures bottom web	Gas barrier level	Overall gauge (μm)
PA/tie/PE/tie/PA/tie/PE	Medium	110 - 170
PA/tie/PA/EVOH/PA/tie/PE	High	110 - 170
PP/tie/PA/EVOH/PA/tie/PE	High	110 - 170
PP/tie/PE/tie/PA/EVOH/PA/tie/PE	High	110 - 170
7 and 9-layer coextruded structures top lidding film	**Gas barrier level**	**Overall gauge (μm)**
PET/tie/PE/tie/PA/EVOH/PA/tie/mLLDPE	High	70 - 110
PET/ink/adh/PE/tie/PA/tie/PE - laminated	Medium	70 - 110
PET/ink/adh/PE/tie/EVOH/tie/PE - laminated	High	70 - 110
OPA/adh/PE - laminated	Medium	70 - 110
OPA/adh/PE/tie/EVOH/tie/PE - laminated	High	70 - 110
PETmet/adh/PE/tie/EVOH/tie/PE [4] -laminated	High	70 - 110
PET-PVDC/adh/PE - coated and laminated	High	70 - 110

[4] PETmet - metalized PET used as outer layer.

Table 5. Examples of MAP structures

Rigid tray	Gas barrier level	Overall gauge (μm)
PVC/adh/PE	Low	200 - 300
APET	Medium	200 - 300
APET/tie/PE	Medium	200 - 300
APET/tie/EVOH/tie/PE	High	200 - 300
PVC/adh/PE/tie/EVOH/PE	High	200 - 300
PS/adh/PE/tie/EVOH/PE	High	200 - 300
EPS/adh/PE/tie/EVOH/PE	High	200 - 300
PP/tie/EVOH/tie/PP	High	200 - 300
APETmet/adh/PE	High	200 - 300
Top lidding film structures	**Barrier**	**Overall gauge (μm)**
PE/tie/PA/tie/PE	Medium	70 - 100
OPA/adh/PE	Medium	70 - 100
PET/adh/PE	Medium	70 - 100
PET/adh/PE/tie/PA/tie/PE	Medium	70 - 100
PA/tie/EVOH/tie/PE	High	70 - 100
OPA/adh/PE/tie/EVOH/tie/PE	High	70 - 100
PET/adh/PE/tie/EVOH/tie/PE	High	70 - 100
Rigid tray	**Gas barrier level**	**Overall gauge (μm)**
PET-PVDC/adh/PE	High	70 - 100
PETmet/adh/PE	High	70 - 100
PET/tie/PE/tie/PA/EVOH/PA/tie/PE	High	70 - 100

Table 6. Examples of retort pouch structures

Laminated film structures	Gas barrier level	Overall gauge (μm)
PET/ink/adh/Al/adh/PP	Medium	100 - 150
PET/ink/adh/Al/adh/PP/tie/PA/tie/PP	High	100 - 150
PET/ink/adh/Al/adh/PET/adh/PP	High	100 - 150
PET/ink/adh/Al/adh/OPA/adh/PP	High	100 - 150
PET/ink/adh/OPA/adh/Al/adh/PP	High	100 - 150
PET/ink/adh/PET-SiOx/adh/OPA/adh/PP	High	100 - 150

Examples of retort pouches include multilayer structures such as: PETink/laminating adhesive/Al//PP, PETink/laminating adhesive/PET/PP, PET/PVDC/PP and PET-SiOx/PP.

Packaging Barrier Properties

The total oxygen available within the package directly affects the quality and the shelf life of red meat. The total oxygen available for oxidative reactions and microbial growth comprises: (i) the residual oxygen left inside the packaging, including the molecules dissolved in the product and remaining in the headspace; and (ii) the molecules that are allowed to enter the packaging by permeation, micro holes and leaks. Therefore, one of the most important features of packaging related to beef preservation is the gas barrier.

The oxygen barrier of packaging films is expressed in terms of oxygen transmission rate (O_2TR). Table 7 shows typical values of O_2TR for films used for red meat packaging.

For modified atmosphere packaging, carbon dioxide barrier is equally important. Although the CO_2 transmission rate is not commonly measured and published, it can be estimated as 4 to 5 times higher than the measured O_2TR.

The effect of product moisture and environmental humidity over hydrophilic polymers gas transmission rates, such as polyamides and EVOH is another factor of utmost importance to be considered throughout the process of designing a gas barrier packaging (Sarantopoulos et al., 2017).

Table 7. Oxygen transmission rate of multilayer films for red meat applications

Multilayer structures	O_2TR* mL(STP).m^{-2}.day^{-1}
Al laminated films	< 0.05
PET–SiOx or OPA–SiOx laminated films	0.3 – 1.0
EVOH coextruded films	0.5 - 15
PVDC coextruded films	10 - 30
OPA coextruded films	30 -150
Plasticized stretch PVC and polyolefic films	10,000 – 15,000

Measured at 25°C and 1 atm.

The ratio between the packaging area and the amount of product must also be considered. The greater the ratio, the greater the film packaging barrier must be. Therefore, smaller portions imply in the use of materials with lower gas transmission rates or the use of thicker barrier layers to achieve a similar shelf life of bigger product portions. Otherwise shorter shelf lives must be defined for smaller portions.

FINAL CONSIDERATIONS

Red meat preservation is a complex matter related to packaging technology and material specifications. Each category of product and market has its specificities in relation to packaging, as was presented in this chapter.

The evolution of beef packaging will pursue the development of convenient and functional packages that ensure safety and traceability; that respond to the demand of sustainability throughout the value chain; and that is flexible to address customized experience meeting lifestyle, entertainment and interactivity requirements. In this scenario, active and intelligent packaging will become continuously more relevant.

Active packaging refers to a series of technologies that allow close interaction between the packaging and its components with the food either by direct contact or by a gaseous headspace. This interaction aims to assure quality and safety throughout specified shelf life. Among these technologies, oxygen scavengers, off-odor scavengers, antimicrobial packages, CO_2 emitters, heat susceptors and pressure release valves (the two latter ones used for microwave cooking) are of interest to flesh products. Oxygen scavengers are by far the most commercially important category of active packaging. They are capable of reducing oxygen levels to less than 0.01%, which is much lower than the typical 0.3–3.0% residual oxygen levels achievable by MAP (Day, 2003). Oxygen scavengers can be used alone or in combination with MAP.

Intelligent packaging monitors and communicates information about the content and surrounding environment to the consumer, retailer and/or

Packaging of Red Meat 221

manufacturer. These technologies usually rely on devices attached to the package such as tags or labels. Indicators of time-temperature, temperature-location, time of use, gases (e.g., O_2 or CO_2), microorganisms, freshness indicators and tracking devices among many others can be used for flesh product packaging. These technologies are associated with chemical or physical sensors, radio-frequency identification (RFID), near field communication (NFC) devices, printed electronics, bidimensional bars codes (e.g., QR Code- quick response code) and information and communication technologies (ICT).

Regarding the future of active and intelligent packaging, it is important for these technologies not to cause concern regarding food safety, not to affect product functionality or its organoleptic features and not to aggregate unintended environmental footprint. It should be mentioned that they should also meet legal requirements. Implementing these systems will require efforts towards education, acceptance and trust among consumers and retailers.

ACKNOWLEDGMENT

Special thanks to Mr. Kleber Brunelli for his generous technical support in collecting and characterizing multilayer structures used for fresh and processed beef packaging.

REFERENCES

Day, B. P. F. Active packaging. In: Coles, R., McDowell, D., Kirkan, M. J. *Food Packaging Technology.* Blackweel Publishing, 2003, p. 282 – 302.

Gazalli, H., Malik, A. H., Jalal, H., Afshan, S. Mir, A. Ashraf, H. Packaging of meat. *International Journal of Food Nutrition and Safety*, 4(2), p. 70-80, 2013.

Luzardo, S., Woerner, D. R., Geornaras, I., Engle, T. E., Delmore, R. J., Hess, A. M., Belk, K. E. Effect of packaging during storage time on retail display shelf life of longissimus muscle from two different beef production systems. *J Anim Sci* 94 (6), p. 2624-2636, 2016.

Morris, B. *The Science and Technology of Flexible Packaging*. Elsevier Inc. – Plastics Design Library, Oxford, 2017, 728 p.

Mullan, M., Mcdowell, D. Modified atmosphere packaging. In: Coles, R., McDowell, D., Kirkan, M. J. *Food Packaging Technology*. Blackwell Publishing, 2003, p. 303 – 339.

NAMI - North American Meat Institute. Case-Ready Meats Modified Atmosphere Packaging. *North American Meat Institute*, January 2015, Washington, D.C., USA. Available at: <http://www.meatinstitute.org>. Access: July, 30th, 2017.

Robertson, G. L. *Food Packaging – Principles and Practice*, CRC Press, Taylor & Francis Group, Florida, 2013, 686 p.

Rodrigues, J. B. M., Sarantopoulos, C. I. G. L., Bromberg, R. Andrade, J. C., Brunelli, K., Miyagusku, L., Marquezini, M. G., Yamada, E. A. Evaluation of the effectiveness of non-irradiated and chlorine free packaging for fresh beef preservation. *Meat Science*, 125, p. 30-36, 2017a.

Rodrigues, J. B. M., Brunelli, K., Sarantopoulos C. I. G. L., Oliveira, L. M. Barrier properties of EVOH and polyamide for fresh beef preservation. *Polímeros Ciência e Tecnologia*, 27, 2017b.

Jin, S. K., Kim, I. S., Song, Y. M., Kim, D. H., Lee, C. Y., Hur, I. C., Park, J., Kang, S. N., Hur, S. J. Effect of packaging methods on quality characteristics of low-grade beef during aging at 16C. *Journal of Food Processing and Preservation*, 37 (6), p. 1111 – 1118, 2013.

Sarantopoulos, C. I. G. L., Alves, R. M. V., Coltro, L., Padula, M., Teixeira, F. G., Moreira, C. Q. Propriedades de barreira. In: Sarantopoulos, C. I. G. L., TEIXIERA, F. G. *Embalagens Plásticas Flexíveis: principais polímeros e avaliação de propriedades*. CETEA/ITAL, Campinas, Brasil, 2017, p. 313 - 358. (Barrier properties. In: Sarantopoulos, C. I. G. L., Teixiera, F. G. *Flexible*

Plastic Packaging: main polymers and properties evaluation. CETEA/ITAL, Campinas, Brazil, 2017, p. 313 - 358).

Scetar, M., Kurek, M. Galie, K. Trends in meat products packaging – a review. *Croatian Journal of Food Science and Technology,* 2 (1), p. 32 – 48, 2010.

Toldra, F. *Handbook of Meat Processing.* John Wiley & Sons, Inc., Iowa, 2010, 560 p.

Zhou, G. H., Xu, X. L., Liu, Y. Preservation technologies for fresh meat – a review. *Meat Science,* 86, p. 119 – 128, 2010.

In: Beef: Production and Management Practices ISBN: 978-1-53613-254-0
Editor: Nelson Roberto Furquim   © 2018 Nova Science Publishers, Inc.

Chapter 9

PRE-SLAUGHTER CATTLE MANAGEMENT IN BRAZIL: ANIMAL WELFARE AND MEAT QUALITY

Bruna Domeneghetti Smaniotto[*] and Bruno Lala
Department of Economy, Sociology and Technology,
São Paulo State University, Botucatu, SP, Brazil

ABSTRACT

The opening of new market niches for Brazil after the sanitary-related events that took place during the 1990s and involved herds from large exporting countries, has boosted participation of Brazil in the international beef trade. Consequently, it was necessary for the country to meet the requirements imposed by the import markets on the quality of the final product, health aspects and animal welfare standards. Given that the national cattle population as a whole has genetic characteristics of the Nellore cattle, which are more reactive than those of the *Bos taurus* species, it is essential to carry out the steps that make up the pre-slaughter

---

[*] Corresponding Author Email: brunadsmaniotto@gmail.com.

management process so that it fits better with the characteristics of production scenario, in order to reduce the unnecessary stress that results in greater occurrence of bruising, metabolic depletion and the DFD (dark, firm and dry) effect in meat, which result in direct and indirect losses for all parties involved in the productive chain. The negative effects generated by the bruises are the result of the high final pH of the meat and lower yield of the carcass and cuts due to the need to remove and discard the affected tissues, mostly located in the most valued and widely marketed parts of the carcass, like tenderloin, filet and rump. Research carried out in the states with the greatest participation in the production of Brazilian beef reported a high occurrence of bruises in the carcasses, which is an evidence of the fact that there are some flaws in the procedures applied during the pre-slaughter process and that there is still a gap in the knowledge of rational management process that would be adequate for these species of cattle. Training and awareness of those who are involved in the beef production chain regarding the importance of pre-slaughter management in animal welfare, the quality of the final product and as preventive factor against production losses should be carried out continuously. Currently, various initiatives are being developed in Brazil in order to warrant a higher level of welfare in cattle, such as those developed by Ministry of Agriculture, Livestock and Food Supply (MAPA), Good Agricultural Practices Program promoted by the Brazilian Agricultural Research Corporation (EMBRAPA) and STEPS — Humanitarian Slaughter, developed by WSPA — World Society for Animal Protection, so that everyone involved in pre-slaughter management could be made aware of and trained according to the needs and behaviors of these animals, developing production and management techniques that, in addition to serving economic interests, would propose greater welfare, less occurrence of DFD meat and increased profitability of the whole sector.

Keywords: management, meat, Nellore, quality, welfare

INTRODUCTION

The opening of new market niches for Brazil after the sanitary events that involved herds of large exporting countries during the 1990s has boosted Brazilian participation in the international beef trade. As a result, the country had to meet the requirements imposed by the import markets

Pre-Slaughter Cattle Management in Brazil 227

on the quality of the final product, health aspects and animal welfare standards (Sanguinet et al., 2013).

Today, the concern for animal welfare is growing worldwide. The demands for ethically produced meat from animals raised and slaughtered in a humane system has not been a concern of just the importing markets, but also of the consumers who have begun to care about the quality of food they buy and consume (da Silva Braga et al., 2014; Vasconcelos Queiroz et al., 2014).

In spite of all the transformations that occurred to the Brazilian beef cattle during the last decade (Moreira, 2014), a significant number of people still see animals as meat-producing machines and ignore many of the psychological and behavioral needs of these animals (Paranhos da Costa et al., 2002).

This fact directly influences the quality of the meat and the profitability of the sector, since it reduces production and quality of the products, causing noncompliance with the requirements of ethical and sustainable production that consider animals to be sentient beings, able to feel pain, different feelings and sentiments (Benez and Polizel Neto, 2015).

The knowledge and importance of the needs and behavior of bovines must be adopted by all those involved in the meat production chain in order to define and implement adequate production and management techniques that, would provide greater welfare to the animals and quality of the final product (Ludtke et al., 2012a).

However, adopting this knowledge in a working routine is still a great challenge for most Brazilian farms, mainly because of the people's reluctance to replace old traditions of rude and stressful handling with those that are calmer and more generous, not to mention any extra investments needed to train the employees and the lack of needed infrastructure at the facilities.

In this chapter we will discuss the stages of pre-slaughter cattle handling in Brazil, looking at the characteristics pertinent to the Brazilian livestock scenario and highlighting factors that interfere with animal welfare, the occurrence of bruises and firm, dry and dark meat (DFD — dark, firm and dry).

ANIMAL WELFARE AND STRESS

Among the various definitions attributed to the term "welfare," the most commonly, use is that which describes as the state of an individual in relation to his or her attempts to adapt to the environment (Broom, 1986). From this angle, any stress agent is capable of altering the homeostasis of an organism and compromise the quality of the meat (Ludtke et al., 2012a; Pighin et al., 2015).

Stress is the main parameter used to evaluate animal welfare, since in this situation the organism triggers a reaction called "alarm response," when physiological, biochemical and behavioral changes occur, which can then be analyzed in the fluids, muscles and the carcass (Ludtke et al., 2012b).

Among physiological responses to stress, we can mention the negative energy balance, muscle glycogen consumption, degradation of protein, dehydration (Suharyanto, 2015), and the secretion of hormones, such as cortisol, adrenaline and noradrenaline, which has as one of its effects the increase in blood glucose concentrations (Pighin et al., 2015).

The hematomas observed during the postmortem evaluation are important indicators of stress and its presence indicates that there were failures during some stages of pre-slaughter handling, thus compromising animal welfare and consequently the meat quality (Barreto, 2014).

MEAT QUALITY

The increase in the occurrence of DFD meat and the high rates of bruising have placed Brazil in the position of being a major producer of raw materials with low added value (Costa, 2013; Gondim, 2013).

BRUISING

Studies have shown an incidence of hematomas in 94.3% of bovine carcasses slaughtered in Pantanal (Andrade et al., 2004), 74.6% (Hensi et al., 2014) and 42.4% (Polizel Neto et al., 2015) in Mato Grosso, contacting the existence of failures in the pre-slaughter of these animals (Paranhos da Costa et al., 2012).

As for the location of the hematomas, they are more likely to occur in the hindquarters, where the most valued parts of the carcass that are marketed raw can be found, such as loin, filet steak, flat cut and rump (de Sousa Pereira et al., 2017; Mendonça et al., 2016; Sornas et al., 2016; Petroni et al., 2013). Hensi et al. (2014), observed hematomas in flat cut (19.56%), rump (18.98%) and rump skirt (13.19%).

The occurrence of bruising is not restricted to hindquarters. Data obtained by Pellecchia (2014) show that besides the rear part (36.15%), they also occur frequently in the ribs (49.02%), the front (23.82%) and the loin (14.59%). The incidence of bruising in these regions is mainly due to the trauma of these parts caused by shocking with the facilities during loading, transportation and unloading stages.

Carcass bruising has multiple origins, and may occur at any stage of pre-slaughter handling, from loading at the rural facility to desensitization in the refrigerator (Morais, 2012). Regardless this, they will always originate from painful situations, serving as an excellent parameter for assessing animal welfare, since its presence indicates that there were faults in one or more stages of this process (Petroni et al., 2013; Benez and Polizel Neto, 2015).

The incidence of hematomas in the carcasses depends on the way the animals are handling during the pre-slaughter (Morais, 2012; Petroni et al., 2013) and of the characteristics pertinent to the animal such as race, reactivity, presence of horns, age and sexual condition (Weeks, Mcnally and Warris, 2002; Pellecchia, 2014; Mendonça et al., 2016).

The Brazilian herd is composed of 215.20 million heads (IBGE, 2015), with a predominance of Zebu breeds (*Bos indicus*), represented by Nellore breed that, although composed of more rustic animals and better adapted to climate and the Brazilian breeding systems, are also more reactive than those of the Taurus breed (*Bos taurus*) (Voisinet et al., 1997).

That characteristic of the Nellore breed, coupled with management errors, making them more agitated and aggressive, favoring the emergence of fights, slips, falls, head shocks and the risk of accidents as a result of their throws against fences and their jumping one over the other, leads to contusion, bruising and the consumption of glycogen in muscles (Ludtke et al., 2012a).

Mendonça et al. (2016) reported a higher occurrence of hematomas in zebu cattle (79.7%) than in Taurus breed, as well as 65.1% of total carcass damage, 131.7% in ribs and 132.8% in the front part of horn animals compared to the hornless ones.

Sexual condition also influences the occurrence of bruising. It is more frequent in females due to the physical constitution of them being smaller when compared to the males, with regard to the percentage of muscle tissue and fat cover (Weeks, Mcnally and Warris, 2002), and because they present a much bonier constitution in the regions of ribs, ischial spine and thighs. This was confirmed by studies that showed at least one hematoma in 57.48% (de Sousa Pereira et al., 2013), 83.8% (Pellecchia et al., 2014) and 18% (Sornas et al., 2016) of female carcasses.

The age of the animal is another influencing factor in the occurrence of bruises in the carcasses, and the occurrence increases with age (Weeks, Mcnally and Warner, 2002).

In the Brazilian cattle ranch, the cows remain for a longer period within the production system and sent to slaughterhouses when they are already at an advanced age, whit reproductive problems, or simply because of commercial decisions related to the livestock cycle (Strappini et al., 2009).

Another point is the direct relation between the occurrence of hematomas in the carcasses and greater transportation distances (Pellecchia, 2014). In studies carried out by Petroni et al. (2013), Vimiso and Muchenje (2013) and Moreira et al. (2014), which evaluated different transportation distances and their relation with the contusion index for bovine carcasses, a higher occurrence of these lesions with an increase in traveled distance was observed.

The need for removal and disposal of parts affected by hematomas brings considerable losses to the industry due to the lower yield of meat cuts and depreciation of carcasses. In Brazil, for every two slaughtered bovines, one has at least one hematoma on average. Each of these lesions, when removed, results in losses of approximately 400 to 600 g of meat per animal (IBCT, 2016).

Based on this information, Petroni et al. (2013) observed losses of approximately 117.3 g per animal in the thigh area alone. Considering the average slaughter of 19.800 cattle and the value of one arroba (approx. 14.7 kg) for the year 2013 (US$ 43.94), the monthly loss with only the toilet of this cut in this year was US$ 6.802,79.

Besides the drawbacks linked to lower yield of the carcass and the cuts because of bruising, there is a close relationship between these and the pH of the meat. Research confirmed this information, where carcasses whit greatest number of hematomas presented pH values above 5.9 (Vimiso & Muchenje, 2013, Pellecchia, 2014).

DFD MEAT

Dry, firm and dark meat (DFD) with high pH (≥ 6.0) is considered the main beef quality problem in Brazil because it generates considerable economic losses due to the high rejection index by part of the consumers at the time of purchase and due to these types of meat failing to reach the most demanding import markets (Ludtke et al., 2012b).

Meat color represents the main attribute evaluated by the consumer at the time of purchase, and if the color is too dark, the other sensory attributes such as softness and taste become insignificant (Dunne et al., 2011). The tenderness of meat represents the most desirable quality attribute associated with consumer satisfaction and, consequently, higher consumption (Malheiros, 2014).

When animals experience aversive events that result in stress and fatigue, the concentration of muscle glycogen tends to decrease. Thus, lactic acid production is compromised by a decrease in postmortem pH index, which results in lower quality meat due to incomplete progression of muscle acidity, producing meat with shorter shelf life, greater water retention capacity, less softness and taste (Petroni et al., 2013; Vimiso and Muchenje, 2013).

The depletion of muscle glycogen reserves and final pH of meat can be used to evaluate physiological and biochemical responses of the organism to stress situations. This is because adequate concentrations of muscle glycogen at the time of slaughter are indispensable for sufficient production of lactic acid and correct pH decline during transformation of muscle into meat (Suharyanto, 2015).

The final pH of meat is a direct function of that concentration. When animals are in stress situations, depletion of muscle glycogen reserves, fat and protein degradation occurs, impeding the drop in postmortem pH and the physical and chemical aspects of meat (Jimenez Filho, 2012).

It should be noticed that import markets, such as the European Union, require meat to have a pH between 5.5 and 5.8, with carcasses with a pH above these values (between 5.8 and 6.0) losing their economic value because they are destined to less demanding markets (Benez and Polizel Neto, 2015).

PRE-SLAUGHTER HANDLING IN BRAZIL

Pre-slaughter handling encompasses all the events that are necessary prior to slaughter. The events that comprise this handling process that start from the separation of the animals in the property, the loading, transportation, unloading and handling in the refrigerator are extrinsic factors capable of influencing the physiological and behavioral responses, as well as the quality of meat (Frimpong et al., 2014).

During the stages of this handling process, animals are constantly in contact with stress factors, which range from the removal of them from their familiar environment to food deprivation, noises, vibrations and the presence of unknown experiences and locations (Leite et al., 2015).

Errors during this handling period, which include overcrowding, lot mixing, shouting, nudging and electric shocks, promote changes in the physiological state of the organism resulting in imbalance of the postmortem metabolism, which manifests itself mainly in the speed of glycolysis and the degree of muscle acidity, increasing the occurrence of DFD meat (Barreto, 2014; Bass et al., 2014).

Before initiated pre-slaughter handling, one should perform the planning, in order to provide better efficiency and reduce the stress of animals and managers. The animals should not stay for long periods in the handling corral and / or in the truck awaiting loading/unloading, provision of documents or equipment and maintenance of facilities, as that increases the risk of accidents and stress (Benez and Polizel Neto, 2015).

It is advisable, then, to use holding pens (Figure 1), which are connecting links between the breeding sites of the animals (pasture) and the handling corral. This because the partition structures of the handling corral, like syringe or cattle crush, accommodates small groups of animals. Moreover, many animals that are waiting for the arrival of the truck for loading should also remain in the chutes until they are shipped (Quintiliano, Páscoa and Paranhos da Costa, 2014; Quintiliano and Paranhos da Costa, 2006).

In the loading of the animals must be respected the load capacity of the vehicle and avoided mixing of animals from different lots and categories in order to prevent fights that result in contusion and bruising of the carcasses (Paranhos da Costa, Spironelli and Quintiliano, 2014).

The most stressful stage of pre-slaughter management is transportation (Ludtke et al., 2012a). During this stage, cattle are exposed to situations that are often unknown to them, such as the actual movement of the vehicle, accelerations and braking, changes in temperature, deprivation of food and water, as well as unknown noises and odors (Frimpong et al.; 2014 Pellecchia, 2014; Franco et al., 2015).

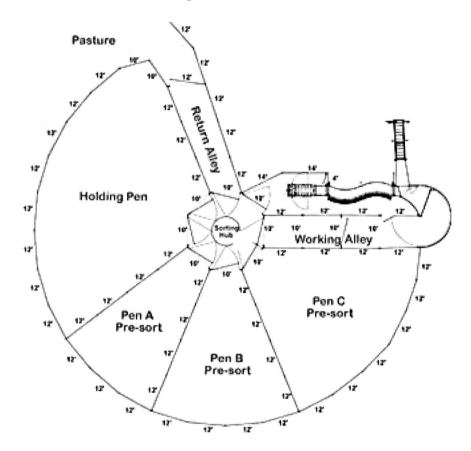

Figure 1. Sample Corral Design. Hi Hog Farm & Ranch Equipment Ltd, USA.

These situations associated with poor transportation conditions cause physiological changes that increase the chances of bruising and risk of excessive stress, which may result in metabolic depletion and even death of the animal (Huertas et al., 2010).

In Brazil, the most commonly used means of transportation of cattle for slaughter is road transportation, carried out by large trucks, carts, double-deck trucks and, less frequently, articulated vehicles with two load compartments called as "Romeo and Juliet" (Franco, 2013), although in some places of the country, river transportation is also common (Andrade, Silva, Roça, 2009).

Pre-Slaughter Cattle Management in Brazil 235

Cattle breeding in the Pantanal area (central west of the country) in extensive system is composed almost exclusively of native pastures that at certain times of the year get flooded by the waters of Paraguay River Basin. Cattle breeding in the Pantanal area (central west of the country) in extensive system is composed almost exclusively of native pastures that at certain times of the year get flooded by the waters of Paraguay River Basin. Due to this characteristic, the flow of production by land is affected and, therefore, the transport of cattle carried out by river, through the so-called corral boats (Andrade et al., 2004).

In the most distant and inaccessible farms in the Pantanal, cowboys take the animals destined for slaughter until the river or road boarding centers. Although river transportation for cattle is of extreme importance in the Pantanal area production, this type of transportation compromises animal welfare by prolonging the handling period (human-animal interaction), transportation and fasting, causing metabolic exhaustion, greater occurrence of lesions and even death of animals (Andrade, Silva and Roça, 2009).

According to Andrade et al. (2007), the river transportation of cattle is a private service, with travel periods that range from one to seven days, during which time the animals receive neither water nor food. In his research, the same author noted a mortality rate of 0.62% for cattle under those transportation conditions.

The results of researches have confirmed the occurrence of at least one lesion in 94.3% (Andrade et al., 2004) and 88.5% (Andrade, Silva and Roça, 2009) of the carcasses pertaining to cattle transported by river.

Vehicle/cart models for road transportation of cattle can also influence the welfare of animals and quality of the meat. Having evaluated different models of vehicles/carts, Franco (2013) concluded that although the "double-deck" model and articulated vehicles with two load compartments (Romeo and Juliet) provide lower freight cost due to their higher load capacity, they also interfere with welfare of the animals and compromise meat quality.

In addition, due to the vast territorial extension, Brazilian beef cattle farms are located, in most cases, far away from slaughterhouses, causing stress to the animals transported over long distances (Barducci et al., 2015).

Long distances are an important source of physical stress to the animals, because during transportation they spend a lot of energy trying to remain balanced within the moving vehicle for a long time. This results in an increase in the occurrence of hematomas, in addition to the depletion of muscle glycogen, leading to an anomalous fall in postmortem pH with significant effects on the quality of the meat (Roça, 2001; Morais, 2012; Schwartzkopf-Genswein et al., 2012, Velarde and Dalmau, 2012).

An example of the evolution of pH in carcasses as a result of different transportation distances can be found in the study of Joaquim (2002), where pH values above 6.0 were observed in 26.6% of animals transported for distances over 300 km.

In addition, another potential effect on animal well-being is the conditions of the roads used to transport the cattle. Of the total of 1.72 million kilometers of roads in Brazil, 12.4% possess paving, with the best highways being located in the Southeast and South regions, and there are many of them in an advanced state of deterioration in the North, Midwest and Northeast areas (IBCT, 2016; Benez and Polizel Neto, 2015).

Therefore, for being a stressful experience for animals, passing through the transportation phase in shorter periods would reduce the duration of this experience that is so painful for animals (Cockram, 2007).

In Brazil, Resolution No. 675 by the National Transit Council (CONTRAM), dated June 2017 and the report titled Animal Well-Being in Maritime or River Live Cattle Transportation (BRASIL, 2017; IBCT, 2016) stand out the importance of the care during by river, maritime and road transportation of production animals and those of economic interest.

According to the National Humanitarian Slaughter Program — STEPS, the unloading of the animals shall occur as soon as arrived at the slaughterhouse. The distance between the truck and the landing ramp shall not permit the formation of spaces capable of hindering the passage of animals and causing injuries and fractures. If there are animals lying in the truck compartment, it is necessary to lift them before starting the unloading (Paranhos da Costa, Quintiliano and Tseimazides, 2014).

An important point to consider is the correct opening of the cargo compartments and their sliding gates, starting with the rear compartment (the closest to the landing stage), and when only two or three animals remain there, the next one should be opened. This prevents two or more animals from attempting to leave at the same time, which would lead to the collision of the animal bodies with the sides of the vehicle, and avoid trauma to the thick rib and back regions of the animals, thus affecting the meat cuts from these parts (Paranhos da Costa, Spironelli and Quintiliano, 2014).

After unloading, the animals should be handled calmly using a flag and directed to the waiting pens, where they will remain at rest, fasting regime and water diet for a maximum of 24 hours until the moment of slaughter (Ludtke et al., 2012a).

CONCLUSION

Pre-slaughter handling is indispensable for the meat production chain, but when performed improperly it has negative impacts on animal welfare and increases the probability of the occurrence of hematomas, metabolic depletion and DFD meat, which result in direct and indirect losses for all those involved in the production chain (Barreto, 2014; Bass et al., 2014; Frimpong et al., 2014).

The adoption of conscious pre-slaughter handling measures that comply with animal welfare standards represents a decisive point in improving the quality of the finished product (Franco, 2013; Benez and Polizel Neto, 2015).

It is possible to reduce the unnecessary stress that may occur during pre-slaughter handling when there is knowledge of the natural behavior of the species and the people involved in the beef production chain are trained and aware of the importance of pre-slaughter animal welfare in the finished product and as a preventive factor to production losses.

Several initiatives developed in Brazil aim to guarantee higher levels of welfare for these cattle, such as those developed by the Ministry of

238 Bruna Domeneghetti Smaniotto and Bruno Lala

Agriculture, Livestock and Food Supply (MAPA), Good Agricultural Practices Program promoted by the Brazilian Agricultural Research Corporation (EMBRAPA) and STEPS — Humanitarian Slaughter, developed by WSPA — World Society for Animal Protection.

Therefore, everyone involved in pre-slaughter handling would be conscious and constantly trained based on the needs and behaviors of these animals, in order to define and enforce production and handling techniques that, in addition to serving economic interests, provide better welfare, less occurrence of bruising and DFD meat, and consequently greater profitability of this important production sector.

REFERENCES

Andrade, E. D., Silva, R. A. M. S. and Roça, R. O. (2009). Manejo pré-abate de bovinos de corte no pantanal, Brasil. *Archivos de Zootecnia* [Pre-slaughter management of beef cattle in the Pantanal, Brazil. *Animal Science Records*], 58(222), 301-304.

Andrade, E. N. D., Silva, R. A. M. S., Roça, R. O., Silva, L. A. C. D., Gonçalves, H. C. and Pinheiro, R. S. B. (2008). Ocorrência de lesões em carcaças de bovinos de corte no Pantanal em função do transporte. *Ciência Rural*, 38(7). [Occurrence of lesions in carcasses of beef cattle in the Pantanal due to transport. *Rural Science, 38* (7)].

Andrade, E. N., Ojeda Filho, S., Da Silva, B. S. and Silva, R. A. M. S. (2004). Influência do transporte fluvial em carcaças de bovinos no Pantanal. *Embrapa Pantanal-Comunicado Técnico (INFOTECA-E)*. [Influence of river transport on cattle carcasses in the Pantanal. *Embrapa Pantanal-Technical Communication (INFOTECA-E)*].

Andrade, E. N., Roça, R. O. and Silva, R. A. M. S. (2007). *Incidência de mortalidade em bovinos transportados por via fluvial no Pantanal.* 3ª Mostra Científica de Ciências Agrárias, 11ª Mostra Científica da FMVZ e 14ª Reunião Científica da Fazenda Lageado. Botucatu, SP. [*Incidence of mortality in beef cattle after fluvial transport in Pantanal.* 3rd Scientific Show of Agricultural Sciences, 11th Scientific

Show of FMVZ and 14th Scientific Meeting of Lageado Farm. Botucatu, SP].

Barducci, R. S., Nave Sarti, L. M., Millen, D. D., Putarov, T. C., Ribeiro, F. A., Silva Franzoi, M. C. D., Costa, C. F., Martins, C. L. and Arrigoni, M. B. (2015). Ácidos graxos no desempenho e nas respostas imunológicas de bovinos Nelore confinados. *Pesquisa Agropecuária Brasileira* [Fatty acids on the performance and immune responses of Nellore feedlot cattle. *Brazilian Agricultural Research*], 50:6, 499-509.

Barreto, E. R. L. (2014). *"Qualidade do manejo no frigorífico: efeitos no bem-estar animal e na qualidade da carcaça e da carne."* 70p. Dissertação (Mestre em Zootecnia) ["Quality of handling in the refrigerator: effects on animal welfare and on the quality of the carcass and meat." 70p. Dissertation (Master in Animal Science)] — Universidade Estadual Paulista, Faculdade de Ciência Agrárias e Veterinárias, Jaboticabal.

Bass, P. D., Engle, T. E., Belk, K. E., Chapman, P. L., Archibeque, S. L., Smith, G. C., and Tatum, J. D. (2010). Effects of sex and short-term magnesium supplementation on stress responses and Longissimus muscle quality characteristics of crossbred cattle. *Journal of Animal Science,* 88(1), 349-360.

Benez, F. M. and Polizel Neto, A. (2015). Pre-slaughter of beef cattle - animal welfare - meat quality. In: *Production and management of beef cattle*, Cuiabá, Mato Grosso: KCM, 118-134.

BRASIL, Brasília, resolução nº 675, de 21 de junho de 2017. Transporte de animais de produção ou interesse econômico, esporte, lazer e exposição. *Diário Oficial da União* [BRAZIL, Brasília, resolution no. 675, June 21, 2017. Transportation of animals of production or economic interest, sport, leisure and exhibition. Official Journal of the Union], 120, 52-53.

Broom, D. M. (1986). Indicators of poor welfare. *British Veterinary Journal,* 142(6), 524-526.

Cockram, M. S. (2007). Criteria and potential reasons for maximum journey times for farm animals destined for slaughter. *Applied Animal Behaviour Science,* 106(4), 234-243.

Costa, F. O. (2013). *"Efeitos do tempo de espera em currais de frigorífico no bem-estar e na qualidade da carne de bovinos."* 96p. Dissertação (Mestre em Zootecnia) - Universidade Estadual Paulista, Faculdade de Ciência Agrárias e Veterinárias, Jaboticabal. ["Effects of waiting time on cold stores on the welfare and quality of beef." 96p. Dissertation (Master in Animal Science) – São Paulo State University, College of Agrarian and Veterinary Science, Jaboticabal].

Da Silva Braga, J., Machado, M. F., Borges, T. D., de Souza, M., de Oliveira Souza, A. P. and Molento, C. F. M. (2014). Diagnóstico de Bem-Estar de Bovinos em dois Matadouros Estaduais na Região Sul do Brasil. *Archives of Veterinary Science*, 19(3). [Diagnosis of Bovine Welfare in two State Slaughterhouses in the Southern Region of Brazil. *Archives of Veterinary Science,* 19 (3)].

De Sousa Pereira, L., de Jesus Santos, G. C., de Lira, T. S., Lopes, F. B., Vieira, Í. A., Minharro, S., Ramos, A. T. and Ferreira, J. L. (2017). Influence of pre-slaughter management on the frequency of injuries and characteristics of bovine carcasses from Southern Pará State, Brazil. *Revista Acadêmica: Ciência Animal,* 11(2), 169-178.

Dunne, P. G., Monahan, F. J. and Moloney, A. P. (2011). Current perspectives on the darker beef often reported from extensively-managed cattle: Does physical activity play a significant role? *Livestock Science*, 142(1), 1-22.

Franco, D., Mato, A., Salgado, F. J., López-Pedrouso, M., Carrera, M., Bravo, S., Parrado M., Gallardo, J. M. and Zapata, C. (2015). Tackling proteome changes in the *Longissimus thoracis* bovine muscle in response to pre-slaughter stress. *Journal of Proteomics*, 122, 73-85.

Franco, M. R. (2013). *"Caracterização do transporte rodoviário de bovinos de corte e efeitos no bem-estar animal e na qualidade as carcaças."* 87p. Dissertação (Mestre em Zootecnia) - Universidade Estadual Paulista, Faculdade de Ciência Agrárias e Veterinárias, Jaboticabal. ["Characterization of road transport of beef cattle and effects on animal welfare and quality of carcasses." 87p. Dissertation (Master in Animal Science) – São Paulo State University, College of Agrarian and Veterinary Science, Jaboticabal].

Frimpong, S., Gebresenbet, G., Bobobee, E., Aklaku, E. D. and Hamdu, I. (2014). Effect of transportation and pre-slaughter handling on welfare and meat quality of cattle: case study of Kumasi abattoir, Ghana. *Veterinary Sciences,* 1(3), 174-191.

Gondim, F. (2013). Bioquímica muscular, maciez da carne e melhoramento das raças zebuínas. Revista de Política Agrícola, 22(4), 95-108. [Muscle biochemistry, meat tenderness and breeding zebu breeds. *Journal of Agricultural Policy,* 22 (4), 95-108].

Hensi, P. C., de Souza, A. L. T. M., de Mello, C. A., Schirmmer, M., de Souza Americano, M. M., Lôbo, A. S. M. T., Gomes, C. C. and da Silva Duarte, S. G. (2014). Perdas Diretas Ocasionadas por Hematomas em Carcaças de Bovinos. Blucher Food Science Proceedings [Direct losses caused by hematomas in bovine carcasses. *Blucher Food Science Proceedings*], 1(1), 433-434.

Huertas, S. M., Gil, A. D., Piaggio, J. M. and Van Eerdenburg, F. J. C. M. (2010). Transportation of beef cattle to slaughterhouses and how this relates to animal welfare and carcase bruising in an extensive production system. *Animal Welfare,* 19(3), 281-285.

Instituto Brasileiro De Geografia E Estatistica (IBGE) (2015). *Produção da Pecuária Municipal* [Brazilian Institute of Geography and Statistics (IBGE) (2015). *Production of Municipal Livestock*], Rio de Janeiro, 43, 1-49.

Instituto Brasileiro De Informação Em Ciencia E Tecnologia (IBICT) (2016). Bem-estar animal no transporte marítimo ou fluvial de animais vivos: panorama da atividade no Brasil e na Espanha. Brasília, 49p. [Brazilian Institute of Information in Science and Technology (IBICT) (2016). Animal welfare in the maritime or fluvial transport of live animals: panorama of the activity in Brazil and Spain. Brasília, 49p].

Jimenez Filho, D. L. (2012). Effect of transport on meat quality-review. *Medicina Veterinaria-Recife,* 6(4), 26-31.

Joaquim, C. F. (2002). *"Efeitos da distância de transporte em parâmetros post-mortem de carcaças bovinas."* 70 p. Dissertação (Mestrado em Medicina Veterinária) - Faculdade de Medicina Veterinária e Zootecnia, Universidade Estadual Paulista, Botucatu. ["Effects of transport distance on postmortem parameters of bovine carcasses." 70 p. Dissertation (Master in Veterinary Medicine) - College of Veterinary Medicine and Animal Science, São Paulo State University, Botucatu].

Leite, C. R., de Mattos Nascimento, M. R. B., de Oliveira Santana, D., Guimarães, E. C. and Morais, H. R. (2015) Influência do manejo pré-abate de bovinos na indústria sobre os parâmetros de bem-estar animal e impactos no ph 24 horas *post-mortem*. *Bioscience Journal*. [Influence of handling pre-slaughter cattle industry on the parameters of animal welfare and impacts on pH 24 hours postmortem. *Bioscience Journal*], 31(1), 194-203.

Ludtke, C. B., Ciocca, J. R. P., Dandin, T., Barbalho, P. C., Vilela, J. A. and Ferrarini, C. (2012a). *Abate humanitário de bovinos* [*Humanitarian slaughter of cattle*]. Rio de Janeiro: WSPA.

Ludtke, C. B., Dalla Costa, O. A., de Oliveira Roça, R., Facco Silveira, E. T., Bortoleto Athayde, N., Pereira de Araújo, A., Mello Júnior, A. and Chinellato de Azambuja, N. (2012b). Bem-estar animal no manejo pré-abate e a influência na qualidade da carne suína e nos parâmetros fisilógicos do estresse. *Ciência Rural* [Animal welfare in pre-slaughter management and the influence on pork quality and physiological parameters of stress. *Rural Science*], 42 (3).

Malheiros, J. M. (2014). *"Identificação e quantificação das proteínas miofibrilares, isoformas da cadeia pesada de miosina (MyHC) e o amaciamento da carne de bovinos Nelore (Bos indicus)."* 80p. Dissertação (Mestrado em Genética e Melhoramento Animal) - Universidade Estadual Paulista, Faculdade de Ciência Agrárias e Veterinárias, Jaboticabal. ["Identification and quantification of myofibrillar proteins, myosin heavy chain isoforms (MyHC) and the breakdown of beef Nellore (*Bos indicus*)." 80p. Dissertation (Master in Genetics and Animal Breeding) – São Paulo State University, College of Agrarian and Veterinary Science, Jaboticabal].

Morais, H. R. (2012). *"Contusões e pH de carcaças de bovinos transportados por diferentes distâncias no verão e inverno."* 35p. Dissertação (Mestrado em Ciência Animal) – Faculdade de Medicina

Pre-Slaughter Cattle Management in Brazil 243

Veterinária, Universidade Federal de Uberlândia, Uberlândia, 2012. ["Contusions and pH of bovine carcasses transported by different distances in summer and winter." 35p. Dissertation (Master in Animal Science) - College of Veterinary Medicine, Federal University of Uberlândia, Uberlândia, 2012].

Moreira, P. S. A., Polizel Neto, A., Martins, L. R., Lourenço, F. J., Palhari, C. and Faria, F. F. (2014). Ocorrência de hematomas em carcaças de bovinos transportados por duas distâncias. Revista Brasileira de Saúde e Produção Animal [Occurrence of hematomas in carcasses of cattle transported for two distances. *Brazilian Journal of Animal Health and Production*], 15(3), 689-695.

Paranhos da Costa, J. R., Quintiliano, M. H, Tseimazides, S. P. (2014). *Boas Práticas de Manejo de Transporte*. Jaboticabal, Funep, 58p. [*Good Practices in Transportation Management*. Jaboticabal, Funep, 58p].

Paranhos Da Costa, J. R., Spironelli, A. L. G., Quintiliano, M. H. (2014). *Boas Práticas de Manejo Embarque*. Jaboticabal, Funep, 37p. [*Good Practices in Shipment Management*. Jaboticabal, Funep, 37p].

Paranhos da Costa, M. J. R., Costa e Silva, E. V., Chiquitelli Neto, M. and Rosa, M. S. (2002). Contribuição dos estudos de comportamento de bovinos para implementação de programas de qualidade de carne. Encontro Anual de Etologia [Contribution of bovine behavior studies to the implementation of meat quality programs. *Annual Meeting of Ethology*], 20, 71-89.

Paranhos da Costa, M. J., Huertas, S. M., Gallo, C. and Dalla Costa, O. A. (2012). Strategies to promote farm animal welfare in Latin America and their effects on carcass and meat quality traits. *Meat Science,* 92(3), 221-226.

Pellecchia, A. J. R. (2014). *"Caracterização do risco de hematomas em carcaças bovinas."* 59p. Dissertação (Mestrado em Zootecnia). Faculdade de Ciências Agrárias e Veterinárias, Universidade Estadual Paulista, Jaboticabal ["Characterization of the risk of bruising in bovine carcasses." 59p. Dissertation (Master in Animal Sciences).

College of Agrarian and Veterinary Sciences, São Paulo State University, Jaboticabal].

Petroni, R., Bürger, K. P., Gonçalez, P. O., Marques, R. G. A., Vidal-Martins, A. M. C. and Aguilar, C. E. G. (2013). Ocorrência de contusões em carcaças bovinas em frigorífico. *Revista Brasileira de Saúde E Produção Animal* [Occurrence of bruises on bovine carcasses in a refrigerator. *Brazilian Journal of Animal Health and Production*], 478-484.

Pighin, D. G., Davies, P., Pazos, A. A., Ceconi, I., Cunzolo, S. A., Mendez, D., Buffarini, M. and Grigioni, G. (2015). Biochemical profiles and physicochemical parameters of beef from cattle raised under contrasting feeding systems and pre-slaughter management. *Animal Production Science*, 55(10), 1310-1317.

Polizel Neto, A., Zanco, N., Lolatto, D. C., Moreira, P. S. and Dromboski, T. (2015). Perdas econômicas ocasionadas por lesões em carcaças de bovinos abatidos em matadouro-frigorífico do norte de Mato Grosso. *Pesquisa Veterinária Brasileira* [Economic losses caused by injuries in cattle carcasses slaughtered in a slaughterhouse in the north of Mato Grosso. *Brazilian Veterinary Research*], 324-328.

Quintiliano, M. H. and Paranhos da Costa, M. J. R. (2006). *Manejo Racional De Bovinos De Corte Em Confinamento: Produtividade E Bem Estar Animal* [*Rational management of beef cattle in cattle shed: productivity and animal welfare*]. IV SINEBOV. Seropédica, R. J.

Quintiliano, M. Q., Pascoa, A. G., Paranhos Da Costa, M. J. R. (2014) *Boas práticas de manejo de Curral: Projeto e construção* [*Good practices of management in cattle shed: Design and construction*]. Jaboticabal: Funep, 57p.

Roça, R. O. (2001). Abate humanitário: manejo ante mortem. *Revista TeC Carnes* [Humanitarian slaughter: ante-mortem management. *Revista TeC Carnes*], 3(1), 7-12.

Sanguinet, E. R., Lorenzoni, R. K., Pelegrine, T., Dörr, A. C., Fruet, A. P. B., Klinger, A. C. K. (2013). International market for brazilian beef: an analysis of the concentration indices of exports from 2000 to 2011.

Electronic Journal in Environmental Management, Education and Technology, 11(11), 2389-2398.

Schwartzkopf-Genswein, K. S., Faucitano, L., Dadgar, S., Shand, P., González, L. A. and Crowe, T. G. (2012). Road transport of cattle, swine and poultry in North America and its impact on animal welfare, carcass and meat quality: A review. *Meat Science*, 92(3), 227-243.

Sornas, A. S.; Rossi Junior, P.; Moizes, F. F. (2016). Losses occasioned by injuries in bovine carcass and its economic reflection in the state of Paraná. *Archives of Veterinary Science*, 21(3), 119-130.

Souza Mendonça, F., Zambarda Vaz, R., Silveira Leal, W., Restle, J., Pascoal, L. L., Bitencourt Vaz, M. and Duarte Farias, G. (2016). Genetic group and horns presence in bruises and economic losses in cattle carcasses. *Semina: Ciências Agrárias*, 37(6), 4265-4274.

Strappini, A. C., Metz, J. H. M., Gallo, C. B. and Kemp, B. (2009). Origin and assessment of bruises in beef cattle at slaughter. *Animal*, 3(5), 728-736.

Strappini, A. C., Metz, J. H. M., Gallo, C., Frankena, K., Vargas, R., De Freslon, I. and Kemp, B. (2013). Bruises in culled cows: when, where and how are they inflicted? *Animal*, 7(3), 485-491.

Suharyanto, S. (2015). Metabolic Responses on Transport Stress and the Effect on Meat Characteristics (A Review). *Journal Sain Peternakan Indonesia*, 4(1), 35-42.

Vasconcelos Queiroz, M. L., Barbosa Filho, D., Antonio, J., Albiero, D., de Freitas Brasil, D. and Melo, R. P. (2014). Percepção dos consumidores sobre o bem-estar dos animais de produção em Fortaleza, Ceará). [Perception of consumers on the welfare of production animals in Fortaleza, Ceará]. *Revista Ciência Agronômica*, 45(2).

Velarde, A. and Dalmau, A. (2012). Animal welfare assessment at slaughter in Europe: Moving from inputs to outputs. *Meat Science*, 92(3), 244-251.

Vimiso, P. and Muchenje, V. (2013). A survey on the effect of transport method on bruises, pH and colour of meat from cattle slaughtered at a

South African commercial abattoir. *South African Journal of Animal Science,* 43(1), 105-111.

Voisinet, B. D., Grandin, T., Tatum, J. D., O'connor, S. F. and Struthers, J. J. (1997). Feedlot cattle with calm temperaments have higher average daily gains than cattle with excitable temperaments. *Journal of Animal Science,* 75(4), 892-896.

Weeks, C. A., McNally, P. W. and Warriss, P. D. (2002). Influence of the design of facilities at auction markets and animal handling procedures on bruising in cattle. *Veterinary Record,* 150(24), 743-748.

ABOUT THE EDITOR

Nelson Roberto Furquim
Doctor Assistant Professor I
Mackenzie Presbyterian University (São Paulo, SP – Brazil)
Email: NRFURQUIM@ALUMNI.USP.BR

Nelson Roberto Furquim has a major in Food Engineering from State University of Campinas - UNICAMP (Brazil), MBA in Management from Business School São Paulo and University of Toronto, with a Master's degree in Business Administration from Mackenzie Presbyterian University, PhD from State University of São Paulo (USP). He is currently a Doctor Assistant Professor at Mackenzie Presbyterian University in São Paulo (Brazil), at Center for Applied Social Sciences (Business Administration). He has over 25 years of corporate experience in management positions, with various international companies, dealing with flavors, packages and commodities. He has also authored or co-authored several specialized publications including papers in selected journals and reviews. His main research lines are management, food technology, nutrition, entrepreneurship and innovation.

LIST OF CONTRIBUTORS

Olivia Borges

Nutrition undergraduate student. Mackenzie Presbyterian University. São Paulo, SP – Brazil.

Ana Maria Bridi

Graduated in Agronomy and Master's degree in Food Science (Federal University of Santa Catarina), PhD in Animal Science (Federal University of Rio Grande do Sul) and postdoctoral degree from the State University of Londrina (UEL). She is currently a Professor at the State University of Londrina. She has experience in the area of Animal Science, with emphasis on Meat Science (evaluation of meats and carcasses), working mainly on the following topics: meat quality, alternative foods and pre-slaughter management. He teaches Masters and PhD students in the UEL Animal Science Postgraduate Program, is a leader of the GPAC (Meat Research and Analysis Group) and a tutor in the PET (Tutorial Education Program) Animal Husbandry . Scholarship from MEC/SESU.

250   *List of Contributors*

Ana Cristina Medeiros Moreira Cabral

Assistant Master Teacher I, Mackenzie Presbyterian University. The author has a degree in Nutrition from the São Camilo University Center (1996) and a Master's Degree in Education, Art and History of Culture from Mackenzie Presbyterian University (2006). She has experience in the area of management of food producing units and nutritional marketing, with emphasis on the development of food products and recipes, customer service and supervision of cooking courses for food companies. Currently, the author is a professor at Mackenzie Presbyterian University in the course of graduation in Nutrition teaching Food Technology, Food Microbiology, Food Hygiene and Legislation, Food composition research, working mainly on the following topics: development of new products, food preparation methods and food processing. She supervised internships in collective feeding, nutritional marketing and commercial food establishments.

Ana Karoline Ferreira Ignácio Câmara

PhD Student at University of Campinas, Professor at the Federal University of São João del-Rei. Food Engineer, Specialist in Meat and Derivatives Technology and Master in Food Technology. Currently, is PhD student at University of Campinas and Professor at the Federal University of São João del-Rei. She has experience in the area of Food Technology and Quality Management with emphasis in the area of meat and meat processing, working on the following topics: reformulation of meat products with a focus on fat reduction, modification of the lipid profile and oxidative stability. Since 2011, she works as a professor in graduate's courses in the area of meat technology and derivatives.

Jorge J. Casal

Chemistry from the Universidad de Buenos Aires, Argentina. Researcher in the National Institute of Agricultural. Technology on meat quality. Professor of Meat Technology in the Universidad Nacional Lomas de Zamora and of Sensorial Analysis and Meat Technology in the Moron University. Buenos Aires, Argentina.

Ulysses Cecato

Bachelor of Animal Science and master's degree in Animal Science (Federal University of Santa Maria) and PhD in Animal Science (São Paulo State University). Post-Doctorate at Range Cattle Research & Education Center - Ona - University of Florida. He is currently a Professor at the State University of Maringá. He has experience in the area of Animal Husbandry, with emphasis on Evaluation, Production and Conservation of Forages, Farming/Livestock Integration, working mainly in Ecophysiology and management of forage plants and production of meat and milk to pasture. Member of the Advisory Committee of the Agricultural Sciences area of the Araucária Foundation.

Jacqueline Dias Machado de Melo

Nutrition undergraduate student. Mackenzie Presbyterian University. São Paulo, SP – Brazil.

Rosires Deliza

Senior Researcher, Embrapa Food Technology. She is a Senior Scientist at the Brazilian Agricultural Research Corporation (Embrapa), working on Sensory Evaluation and Consumer Science. She was trained as a Food Engineer and has a Food Science MSc degree both from the UNICAMP, Brazil. She received her PhD in Food Science from the University of Reading, UK. She worked two years at Embrapa Labex Europe, based at INRA, UMR CSGA, Dijon-France. She supervises MSc and PhD students. During her research career, she has authored or co-authored over 160 original papers, book chapters, review articles and technical publications. She received the Scientist of Rio de Janeiro State Award, which is offered to the State Outstanding Scientists.

Mateus Silva Ferreira

BS, Animal Science, State University of Maringá, 2014.
MS, Animal Science, State University of Maringá, 2017.
PhD student, Animal Science, São Paulo State University.

Maria Teresa Esteves Lopes Galvão

PhD student, University of Campinas. She is working on Sensory Evaluation and Consumer Science as a consultant. She graduated on Food Engineering at the University of Campinas (UNICAMP) and has Food Technology MSc and PhD both from UNICAMP. She worked for eight years at the Meat Technology Center at Institute of Food Technology (ITAL) and for 15 years in a Brazilian private food company (SADIA) as a scientific researcher and sensory evaluation manager. During her career she has authored or co-authored more than 20 original papers, book chapters and technical publications.

Pilar Teresa Garcia

Emeritus Investigator in the Institute Food Technology (INTA). Castelar Bs As Argentina. President of "Forum Food Nutrition and Human

List of Contributors

health" (FANUS). Bolsa de Cereales de Buenos Aires. Argentina. Institute of National Institute of Food Technology (INTA) Castelar; Facultad de Agronomia y Ciencias Agroalimentarias. Universidad de Moron. Bs As Argentina; Facultad de Zootecnia. Universidad de Lomas de Zamora. Bs As. Argentina. Researcher in the Institute of National Institute of Food Technology (INTA) Castelar. Professor of Food Analysis and Nutrition. Facultad de Agronomia y Ciencias Agroalimentarias. Universidad de Moron. Bs As Argentina. Professor of Meat Technology. Facultad de Zootecnia. Universidad de Lomas de Zamora. Bs As. Argentina.

Josiane Fonseca Lage

BS, Animal Science, Federal University of Viçosa, 2007.
MS, Animal Science, Federal University of Viçosa, 2009.
PhD, Animal Science, São Paulo State University, 2014.
Research and Development Supervisor at Trouw Nutrition.

Bruno Lala

PhD student, Department of Economy, Sociology and Technology, São Paulo State University, Botucatu, São Paulo, Brazil. Animal Scientist (2010), with a master's degree in Animal Science – Pastures and Forage Farming (2015) in the Graduate Program in Animal Science linked to the Center of Agrarian Sciences, both at the State University of Maringá (UEM, Maringá, Paraná, Brazil). He was a scientific initiation scholarship student by National Council for Scientific and Technological Development (CNPq) and Araucária Foundation (Paraná, Brazil) and had a technical support fellow scholarship by CNPq. He was a substitute professor in the disciplines of Biochemistry, Plant Physiology, Plant Morphology and Applied Zoology for the course of Agronomy at University Center of Maringá (UNICESUMAR, Maringá, Paraná, Brazil), and Animal Production for the course of Agribusiness at São Paulo Technology

College (FATEC, Presidente Prudente, São Paulo, Brazil). He achieved the title *"Best Oral Presentation"* at the 60th International Congress of Meat Science and Technology – ICoMST (2014). He is a PhD student of the Graduate Program in Animal Science at the Veterinary Medicine and Animal Science School, São Paulo State University (UNESP, Botucatu, São Paulo, Brazil) – CNPq Scolarship. He has experience in Animal Science, working mainly in the following subjects: animal production, carcass quality, meat quality, gene expression and oxidative stress.

Nestor N. Latimori

Veterinary surgeon, Facultad de Agronomia y Veterinaria, Universidad Nacional de Rio Cuarto, Cordoba, Argentina. Magister Science in Animal Production. UN Mar del Plata. Buenos Aires. Argentina. Researcher on Meat Production Systems and meat quality and attributes in the National Institute of Agricultural Technologies (INTA) Cordoba, Argentina.

Otávio Rodrigues Machado Neto

Assistant Professor, São Paulo State University – College of Veterinary, Medicine and Animal Science, Botucatu, SP.
Bachelor in Animal Science, Federal Rural University of Rio de Janeiro., 2002.
MSc in Animal Science, Federal University of Lavras, 2008.
DSc in Animal Science, Federal University of Lavras, 2011.

Vinícius Valim Pereira

Bachelor of Animal Science (Federal Rural University of Rio de Janeiro), Master of Forragriculture and Pastures (Federal University of Viçosa), Specialist in Agribusiness with an emphasis on Markets and PhD in Animal Sciences - Pasture and Forage Farming (State University of Maringá). He has experience of working in the area of Forage and Pasture, Project Management for Agribusiness, Dairy and Beef Cattle, Sustainability and Biotechnology. Currently holds post doctorate at the Nucleus of Biochemistry in the area of Biotechnology of Microorganisms Federal University of São João Del-Rei. He is currently a Professor at Pitágoras College - Divinópolis Campus

Marise Aparecida Rodrigues Pollonio

Full Professor of Food Technology, State University of Campinas. Marise Aparecida Rodrigues Pollonio is Food Engineer from State University of Campinas (Brazil), with a Master's degree in Food Science, PhD from State University of Campinas-UNICAMP and post-doctorate from Instituto de Agroquìmica y Tecnología de Alimentos (IATA - CSIC), Valencia, Spain. She is currently a full professor at the State University of Campinas in Food Technology Department and coordinator of Post Graduate Program of Food Technology of Faculty of Food Engineering. She has authored or co-authored over 150 specialized publications including papers in selected periodicals, short papers an Abstracts in international events. She advised 30 thesis and dissertations. Her main research lines are reformulating of meat products and development of healthier strategies to reduce fat, sodium and additives in processed meats, including also study of bioactive compounds.

Ana Paula Possamai

Bachelor of Animal Science (State University of Londrina), Master's Degree and PhD in Animal Science - Production and Nutrition of Ruminants (State University of Maringá). She is currently Professor and Coordinator of the undergraduate courses in Animal Science, Agronomy and Veterinary Medicine of the Faculdades Unidadas do Vale do Araguaia

256 *List of Contributors*

(Mato Grosso, Brazil). She has experience in the area of Animal Science, with emphasis on Animal Nutrition and Feeding, Ruminal Kinetics, Ruminant Production, Meat Quality and Business Quality Management, working mainly on the following topics: nutrition and ruminant production, animal feed, system implantation HACCP, Good Manufacturing Practices and Quality Control of the refrigeration industry.

Liziana Maria Rodrigues

BS, Animal Science, Federal Rural University of Rio de Janeiro, 2008.
MS, Animal Science, Federal Rural University of Rio de Janeiro, 2009.
PhD student, Animal Science, Federal University of Lavras.

José Boaventura M. Rodrigues

Flexible Packaging Account Manager – DuPont do Brasil S.A., Maua Institute of Technology, Sao Paulo, Brazil. Bachelor and MSc in Chemical Engineering Jose Boaventura M. Rodrigues started his carrier as polyolefin process and product R&D engineer. Presently his carrier accounts 30 years experience in various distinctive areas including polyethylene homopolymer and copolymer development for food packaging, agriculture and industrial applications. As 6-Sigma Black Belt, Jose Boaventura has coordinated several projects aiming business strategy deployment and supply chain process integration. As business development manager he coordinated projects for the development of meat multilayer structure packaging. His present research focuses on optimizing food plant efficiency, by evaluating the performance of food packing lines, more specifically adapting the Overall Equipment Effectiveness - OEE technique for small and medium size food plants as an estimator of their financial performance.

List of Contributors

Guilherme Sicca Lopes Sampaio

Veterinarian by the Federal University of Uberlândia, Master in Veterinary Hygiene and Technological Processing of Animal Products (Fluminense Federal University). PhD in Inspection of Products of Animal Origin (São Paulo State University). Postdoctoral fellow at the Faculty of Agronomic Sciences, (São Paulo State University). Researches: Strategies to reduce pathogens and deteriorants in fresh beef. Other activities: Consulting for the meat industry: design and execution of research projects, data analysis (biostatistics) and technical-scientific writing.

Ana M. Sancho

Magister in Statistical Biometry from the Universidad de Buenos Aires (UBA). Reseacher in Systems of Information in the National Institute of Agricultural Technology (INTA), Argentina.

Claire Isabel G. L. Sarantópoulos

Senior Scientist, Center of Packaging Technology of the Food Technology Institute, Brazil. Food Technology Institute, Campinas, Brazil. Senior Scientific Researcher at Food Technology Institute, Brazil, Claire Sarantopoulos brings over 30 years of experience working in packaging in both RD&I and academia. In her role as a packaging researcher, she focuses on studying plastic packages with a special focus on barrier properties, active and intelligent packages, and packed food and beverage preservation. As a sought-after packaging expert, she has recently led four studies on consumer trends related to the packaging of foods and beverages. She also serves as editor in chief of the Brazilian Journal of Food Technology and frequently collaborates with Brazilian Packaging Magazines. Over the course of her career, she has authored or co-authored 44 books and over 50 original research papers on package performance and shelf life of packed foods. She earned her degree as a Food Engineer from the Universidade Estadual de Campinas - UNICAMP, Brazil.

Natália De Luca Silva

Nutrition undergraduate student. Mackenzie Presbyterian University. São Paulo, SP – Brazil.

Paula Louro Silva

Nutrition undergraduate student. Mackenzie Presbyterian University. São Paulo, SP – Brazil.

Bruna Domeneghetti Smaniotto

Teacher Assistant II, São Paulo University (UNIP), Bauru, SP, Brazil. Graduation in Veterinary Medicine by São Paulo University, Bauru Campus (2008-2012). Residency in Veterinary Medicine in the area of Avian Pathology by the Faculty of Veterinary Medicine and Zootechny (FMVZ), of the São Paulo State University (UNESP), Botucatu (2013-2015). Master's Degree in Animal Production by the Graduate Program in Zootechny (FMVZ/UNESP) (2015-2017), with emphasis on beef cattle production, rational management, animal welfare and bovine meat quality.

INDEX

Δ9

Δ9 desaturase, 141, 142

A

absorber, 202
acceptance, 15, 17, 19, 36, 46, 58, 59, 75, 99, 116, 157, 196, 204, 221
accession, 70, 72, 75, 79, 82, 83
accidents, 95, 230, 233
accreditation, 73
active and intelligent packaging, 220, 221
adipocyte, 125, 141, 181, 185, 186, 190
adipogenesis, 180, 181, 182, 186
adolescence, 9, 32
adulteration, 75, 81
aerobic, 196, 203
age, xv, 1, 3, 9, 95, 98, 103, 104, 105, 113, 115, 156, 158, 177, 187, 195, 229, 230
agreements, 70, 82, 94
agribusiness, xii, 7, 70, 72, 82, 89, 165, 252
anaerobic, 196, 203

animal performance, 127, 153, 155, 157, 164, 166, 167, 172
animal production, 153, 154, 155, 156, 157, 164, 166, 167, 252
animal welfare, xiii, 225, 227, 228, 229, 235, 237, 239, 240, 241, 242, 243, 244, 245, 254
animal well-being, 236
antibiotics, 71
appearance, 2, 17, 18, 19, 50, 106, 107, 108, 185, 194, 195, 196, 200, 202, 204, 205, 206, 207, 208, 209, 213
atherogenic, 124, 129, 137
attribute, 17, 18, 36, 59, 60, 97, 99, 100, 101, 107, 115, 197, 231

B

barrier, 73, 194, 198, 199, 201, 202, 204, 205, 206, 207, 208, 209, 212, 213, 214, 215, 216, 217, 218, 219, 220, 222, 254
beef exports, 73, 77, 80, 83
beef hamburger, 3
beef intramuscular fat, 124, 144
beef preservation, 194, 219, 222

260 *Index*

behavior of bovines, 227
bloom, 197, 202
bologna sausages, 36, 37, 38, 40, 41, 42, 44, 48, 49, 50, 51, 52, 53, 54, 55, 56, 57, 58, 59, 60
bone, 95, 115, 116, 170, 203, 205, 207
bovine carcasses, 10, 229, 231, 240, 241, 242, 243, 244
bovine spongiform encephalopathy (BSE), 69, 71, 72, 86, 89
bovine(s), 3, 6, 7, 10, 15, 27, 69, 71, 72, 73, 74, 75, 82, 85, 86, 87, 144, 154, 159, 160, 166, 178, 227, 229, 231, 240, 241, 242, 243, 244, 245, 254
Brachiaria, 153, 154, 157, 167, 173
Brazilian Agricultural Research Corporation, xii, 31, 226, 238, 250
Brazilian beef, 23, 72, 73, 77, 78, 81, 83, 88, 89, 90, 152, 226, 227, 235
Brazilian beef cattle, 152, 227, 235
Brazilian breeding systems, 230
Brazilian farms, 227
Brazilian market, 2, 15, 21, 22, 24, 99
Brazilian system of identification and certificate of bovine and buffalo origin (SISBOV), 7, 27, 70, 71, 72, 73, 74, 75, 76, 77, 78, 79, 80, 81, 82, 83, 84, 85, 86
breed, 125, 127, 141, 144, 147, 148, 149, 150, 168, 184, 187, 195, 229, 230
breeding, 64, 74, 79, 90, 154, 171, 176, 230, 233, 235, 241, 242
bribes, 75
browning, 202, 203, 211
bruises, 226, 227, 230, 244, 245
bubaline, 70, 71, 72, 74, 75, 85
buffalo, 6, 7, 73, 74, 85

C

carbon dioxide, 9, 197, 198, 208, 219
carbon monoxide, 197, 198

carcass, xv, 23, 124, 141, 149, 153, 154, 159, 160, 166, 168, 169, 170, 172, 173, 176, 177, 183, 184, 185, 188, 189, 190, 191, 201, 226, 228, 229, 230, 231, 239, 243, 245, 252
carcass bruising, 229
cardboard, 2, 14, 15, 211
cardiovascular health, 11
cargo compartments, 237
case-ready, 194, 196, 209
casings, 43, 100, 201, 214
cattle, ix, xv, 4, 6, 23, 31, 33, 70, 71, 72, 73, 74, 75, 76, 77, 78, 79, 80, 81, 82, 83, 84, 88, 89, 90, 125, 141, 144, 146, 147, 149, 152, 153, 154, 159, 160, 161, 162, 163, 164, 167, 169, 170, 171, 172, 173, 176, 177, 179, 181, 184, 187, 188, 189, 190, 191, 225, 227, 230, 231, 233, 234, 235, 236, 237, 238, 239, 240, 241, 242, 243, 244, 245, 246, 254
cattle breeding, 90, 154, 171, 235
cattle ranch, 70, 72, 74, 75, 76, 77, 78, 79, 80, 83, 230
central ideas, 76
certifying entity, 74
chain, viii, xii, 4, 7, 11, 24, 30, 36, 38, 61, 65, 69, 70, 71, 72, 73, 74, 75, 76, 78, 80, 82, 83, 85, 86, 87, 89, 90, 91, 117, 125, 126, 132, 141, 162, 172, 188, 193, 199, 200, 220, 226, 227, 237, 242, 253
cholesterol, viii, 11, 66, 104, 123, 124, 126, 129, 130, 138, 144, 145, 147, 149, 150, 166
cholesterol content, 66, 126, 144, 147, 149
cis-9, 126, 162, 163
claim label, 36
clandestine slaughtering, 80, 81
cleaner labels, 39
climate, 154, 164, 184, 230
CO_2, 158, 197, 203, 204, 209, 210, 211, 219, 220, 221
collision, 237

color, 10, 36, 44, 46, 50, 51, 59, 60, 106, 107, 108, 109, 160, 178, 194, 195, 196, 197, 198, 199, 202, 203, 204, 207, 209, 210, 211, 213, 231
comminuted meat products, 37
conditions of the roads, 236
conjugated linoleic acid (CLA), 65, 125, 126, 131, 133, 134, 137, 139, 141, 146, 147, 148, 150, 154, 156, 162, 163, 168, 169, 170, 171
connective tissue, 10, 37, 176, 181, 187
conscious, 237, 238
consumer satisfaction, 97, 108, 114, 115, 231
consumer(s), viii, 4, 5, 8, 13, 14, 15, 17, 18, 22, 23, 24, 35, 58, 63, 69, 71, 72, 73, 91, 93, 94, 96, 97, 98, 99, 100, 102, 103, 104, 105, 106, 107, 108, 109, 110, 111, 112, 113, 114, 115, 116, 117, 118, 119, 120, 121, 125, 146, 160, 172, 185, 187, 193, 194, 195, 197, 200, 201, 202, 203, 204, 209, 210, 212, 220, 221, 227, 231, 245, 250, 251, 254
consumers' perception, 94, 96, 102, 118, 120
consumers' expectation, 96
consumption, 1, 3, 4, 10, 11, 12, 17, 18, 19, 25, 35, 37, 72, 93, 95, 98, 99, 103, 104, 105, 106, 112, 115, 116, 118, 133, 149, 153, 157, 158, 159, 160, 161, 164, 175, 209, 228, 230, 231
consumption of glycogen, 230
contaminants, 4
contaminated beef, 71, 72
contamination problems, 71
CONTRAM, 236
control mechanisms, 82
convenience, 2, 24, 70, 76, 96, 115, 200
corral boats, 235
covers, 10, 49, 201

D

death of animals, 235
decision making, 146
dehydration, 47, 196, 198, 228
demanding markets, 232
deoxymyoglobin, 197, 210
depreciation, 159, 231
desaturase activity, 132, 141, 142, 147
DFD meat, 226, 228, 233, 237, 238
DHA, 57, 126, 135
diet, 20, 24, 31, 37, 67, 94, 95, 98, 102, 103, 104, 105, 112, 115, 119, 120, 124, 125, 126, 127, 129, 130, 131, 132, 133, 135, 136, 137, 138, 140, 141, 144, 145, 146, 148, 153, 158, 161, 163, 173, 178, 181, 182, 183, 184, 186, 237
dietary cholesterol, 144
discoloration, 195, 199, 202
discourse of the collective subject (DCS), 70, 76, 88
drip, 195, 196, 201, 202, 205
dripping, 199, 203, 207
duration, 184, 236

E

economic actors, 70, 76
economic interests, 226, 238
economic losses, 195, 231, 244, 245
effectiveness, 47, 70, 75, 76, 79, 81, 83, 119, 222, 254
electric shocks, 233
embargo, 75, 81
emulsion, 15, 40, 42, 43, 44, 49, 50, 55, 62
energy density of the diet, 184
energy intake, 128, 184
engineers, 77
enzymatic and nutritional indices, 124
enzymatic indices, 124, 126, 137, 140, 142, 143, 145

262 *Index*

enzyme activity, xv, 141
EPA, 56, 57, 126, 135
epigenetic, 176, 189
equipment, 128, 202, 204, 205, 206, 211, 213, 233, 234, 253
ethical, 227
ethically produced meat, 227
ethics, 17, 46, 76, 81, 83, 98
ethylene vinyl alcohol (EVOH), 215, 216, 217, 218, 219, 222
exports, 73, 74, 75, 77, 80, 81, 82, 83, 91, 245
extensive system, 235
extrinsic factors, 196, 232

F

fast-food, 9
fatigue, 232
fatty acid, xv, 12, 36, 37, 38, 40, 45, 55, 56, 57, 58, 62, 63, 64, 65, 66, 67, 124, 125, 126, 128, 129, 131, 134, 136, 141, 144, 145, 146, 147, 148, 149, 150, 153, 154, 156, 161, 162, 163, 166, 167, 168, 169, 171, 172, 173, 181, 185, 186, 239
fatty acid profiles, 38, 147, 166
fatty acid profiles of meat and meat products, 38
fatty acid(s), xv, 12, 36, 37, 38, 40, 45, 55, 56, 57, 58, 61, 62, 63, 64, 65, 66, 67, 124, 125, 126, 128, 129, 131, 134, 136, 141, 144, 145, 146, 147, 148, 149, 150, 153, 154, 156, 161, 162, 163, 166, 167, 168, 169, 171, 172, 173, 181, 185, 186, 239
fatty acids composition, 124, 148, 149
feedlots, 188
females, 160, 230
fetal programming, xvi, 176, 184, 189
film, 2, 45, 201, 202, 203, 205, 206, 207, 208, 209, 213, 216, 217, 218, 220

final product, 164, 188, 225, 227
financial returns, 77
flavor, xv, 18, 19, 24, 36, 37, 39, 41, 46, 59, 60, 63, 64, 94, 96, 99, 106, 107, 108, 110, 118, 153, 160, 178, 185, 195, 196, 199, 210
flowpack(s), 202, 203, 208
focus group(s), 94, 97, 98, 99, 100, 101, 103, 107, 108, 115, 116, 118
food, vii, xi, xii, xiii, 1, 2, 3, 4, 7, 9, 10, 11, 12, 13, 14, 20, 21, 22, 23, 24, 25, 29, 30, 31, 32, 33, 35, 36, 47, 61, 62, 63, 64, 65, 66, 67, 69, 70, 71, 73, 75, 76, 78, 79, 82, 83, 87, 89, 90, 93, 94, 95, 97, 98, 99, 102, 104, 112, 114, 115,116, 117, 118, 119, 120, 121, 125, 133, 146, 148, 152, 154, 158, 160, 173, 184, 193, 213, 220, 221, 222, 223, 226, 227, 233, 235, 238, 241, 247, 249, 250, 251, 253, 254
food deprivation, 233
food industry, xi, 3, 22, 116
food inspection, 70, 71, 76
food preservation, 12
food safety, 4, 7, 36, 69, 70, 71, 78, 90, 221
foodborne diseases, 71
foot-and-mouth disease, 6, 75, 81, 83
forage, 125, 127, 133, 141, 147, 148, 153, 154, 155, 156, 157, 158, 161, 162, 163, 164, 165, 167, 168, 169, 170, 171, 172, 173, 252
fortification, 2, 3
fractures, 236
frankfurter type sausages, viii, 93
freezer burn, 198, 207
freezing, 2, 10, 13, 30, 106, 198, 207
frequency of hamburger consumption, 17, 18, 19
freshness, 197, 202, 221
functional meat products, 37

Index 263

G

gas, 45, 128, 195, 196, 199, 202, 206, 207, 208, 209, 210, 211, 212, 213, 214, 215, 216, 217, 218, 219, 220
genotype, 124, 125, 127, 128, 129, 130, 131, 132, 133, 135, 136, 137, 138, 140, 141, 144, 145, 148, 150, 166, 170, 187
glycolysis, 178, 233
good agricultural practices program, 226, 238
grass, 125, 147, 148, 149, 152, 153, 154, 155, 156, 157, 158, 163, 165, 166, 168, 172

H

hamburger, vii, 1, 2, 3, 15, 17, 18, 19, 20, 21, 22, 23, 24, 106
handling corral, 233
handling techniques, 158, 238
health surveillance, 5
healthiness, xiii, 98, 103, 104, 112, 113, 117, 133
hematomas, 228, 229, 230, 231, 236, 237, 241, 243
herb and spice blends, 36, 40, 41, 43, 49, 50, 51, 53, 55, 58, 59, 60
herbs and spices, 36, 39, 40, 41, 42, 43, 47, 48, 51, 52, 58, 60, 67
herd, 23, 72, 73, 153, 229
heritable, 185
higher antioxidant, 36, 40
humanitarian slaughter, 226, 236, 238, 242, 244
hypertrophy, 125, 141, 177, 179, 181, 182

I

identification, 7, 27, 70, 71, 72, 73, 74, 75, 76, 79, 84, 86, 87, 107, 216, 221, 242
IMF%, 135, 136, 144, 145
implementation, 29, 33, 73, 75, 79, 80, 81, 86, 87, 90, 243
indicators of stress, 228
industrialized foods, 9
industrialized products, 3, 94
information asymmetry, 80
institutional environment, 32, 70, 75, 87
international commercialization, 73
international standards, 79
intramuscular adipocytes, xvi, 181, 186
intramuscular fat, xv, 124, 126, 128, 134, 136, 139, 140, 143, 144, 145, 146, 153, 160, 163, 168, 178, 181, 183, 184, 185, 188, 191
ionomers, 214
ipids, xvi, 125, 136, 148, 156, 168
iron, vii, 1, 2, 3, 8, 9, 10, 12, 15, 20, 21, 23, 24, 31, 32, 37, 47, 197
iron deficiency, 1, 2, 8, 9, 24, 32
isomers, 125, 146, 148, 150, 161, 162, 163, 171

K

Kano Method, 108
Kano modeling, 97
Kano questionnaire, 102
key expressions (KE), 76

L

labeling, 2, 7, 8, 27, 79, 200, 216
lactic acid, 178, 196, 232
landing ramp, 236
lesions, 231, 235, 238

lid, 206, 207, 209, 210

linseed oil, 36, 38, 40, 41, 42, 43, 48, 49, 50, 51, 52, 53, 54, 55, 56, 57, 58, 59, 60, 61, 62, 63, 64, 66

lipid oxidation, 36, 38, 39, 47, 48, 52, 60, 61, 62, 63, 65, 66

live cattle transportation, 236

liver, vii, 1, 2, 3, 10, 15, 16, 18, 19, 20, 22, 23, 33, 67, 186

livestock, 2, 5, 6, 7, 8, 23, 27, 28, 29, 30, 31, 71, 72, 73, 84, 85, 86, 89, 90, 147, 152, 153, 155, 165, 169, 170, 171, 173, 189, 190, 195, 226, 227, 230, 238, 240, 241

livestock scenario, 227

loading, 229, 232, 233

long distances, 236

long-chain n-3 LC-PUFAs, 125

losses, 13, 23, 36, 47, 54, 96, 193, 195, 196, 226, 231, 237, 241, 244, 245

low sodium meat products, 94

M

machines, 205, 227

malnutrition, 3, 9, 176, 177, 183

management, viii, ix, xii, xv, 23, 24, 70, 72, 80, 82, 85, 87, 127, 151, 152, 153, 155, 156, 157, 164, 165, 167, 170, 173, 184, 187, 188, 189, 225, 226, 227, 230, 233, 238, 239, 240, 242, 243, 244, 245, 247, 249, 250, 254

management errors, 230

managers, 233

mandatory, 70, 72, 74, 79, 83

MAPA, 2, 5, 6, 7, 8, 27, 30, 73, 74, 76, 78, 79, 80, 81, 87, 226, 238

marbling, xv, xvi, 124, 142, 149, 153, 160, 178, 181, 182, 183, 187

maritime, 236, 241

markets, 7, 8, 22, 37, 70, 73, 74, 81, 82, 83, 126, 225, 226, 227, 231, 232, 246

masterpacks, 203

maternal malnutrition, 176, 177

meat and meat products, 35, 37, 38, 39, 99, 120

meat batter, 37, 41, 43, 44, 47, 54

meat color, 160, 178, 197, 231

meat composition, 146

meat cuts, 213, 231, 237

meat products, 11, 36, 37, 38, 39, 47, 48, 49, 54, 58, 60, 61, 63, 94, 95, 98, 99, 103, 104, 106, 113, 119, 120, 144, 200, 223, 250, 253

meat quality, xvi, 4, 23, 67, 124, 150, 152, 153, 160, 168, 169, 173, 176, 177, 178, 181, 184, 185, 187, 190, 228, 235, 239, 241, 243, 245, 252, 254

meat-based products, 75

metabolic exhaustion, 235

metmyoglobin, 197, 202, 211

microbial growth, 195, 196, 197, 199, 202, 210, 219

micronutrients, 3

microorganisms, 14, 71, 195, 196, 197, 203, 212, 221

modified atmosphere packaging (MAP), 194, 197, 198, 201, 208, 209, 210, 211, 215, 216, 218, 219, 220, 222

modified lipid profiles, 58

moisture barrier, 198, 202, 208, 215

monitoring systems, 71

monounsaturated fatty acids (MUFAs), 38, 55, 124, 125, 126, 129, 132, 136, 141, 150

MUFA/SFA ratio, 136

muscle acidity, 232, 233

muscle glycogen, 228, 232, 236

muscle glycogen consumption, 228

muscle(s), xvi, 9, 10, 64, 65, 124, 125, 128, 131, 133, 134, 136, 137, 138, 139, 140, 141, 143, 144, 146, 147, 148, 149, 150, 159, 161, 166, 167, 172, 175, 176, 177, 178, 179, 180, 181, 182, 183, 184, 185,

Index

187, 188, 189, 190, 191, 192, 195, 197, 199, 204, 211, 222,228, 230, 232, 233, 236, 239, 240, 241

myogenesis, 180, 181, 182

myoglobin, 10, 161, 195, 196, 197, 198, 199, 202, 204, 210, 211

N

natural antioxidant(s), 36, 39, 63, 64, 66

Nellore, 225, 226, 229, 230, 239, 242

Nellore breed, 229, 230

non-probabilistic, 2, 17, 70, 76

non-tariff barriers, 72

nutrients intake, 3

nutritional, viii, xiii, xv, 1, 2, 3, 4, 8, 9, 11, 12, 13, 15, 18, 20, 21, 22, 24, 27, 29, 32, 35, 37, 38, 40, 58, 61, 66, 79, 82, 94, 104, 108, 109, 115, 117, 123, 124, 126, 130, 133, 135, 136, 138, 139, 145, 146, 149, 152, 157, 164, 165, 173, 175, 176, 181, 182, 183, 184, 188, 189, 249

nutritional composition, 24

nutritional deficiencies, 3

nutritional indices, 124, 135, 138, 139

nutritional information table, 18

nutritional status, 66, 182

nutritional value, 4, 12, 13, 21, 24, 35, 37, 38, 61, 124, 135, 136, 146, 152

O

O_2, 196, 197, 198, 204, 209, 210, 211, 221

O_2TR, 219

obesity, 3, 22, 95, 104, 113, 181

odor, 2, 17, 18, 19, 195, 196, 199, 202, 211, 220

off-flavor, 195, 199, 210, 211

oligopsone, 80, 88

omega 3, 126, 161

opportunistic behavior, 72, 80, 83

origin, 3, 5, 7, 8, 10, 12, 23, 25, 26, 27, 30, 33, 37, 70, 71, 72, 73, 74, 85, 94, 95, 160, 162, 195, 245

overcrowding, 233

overprice, 70, 79, 80

overweight, 3

overwrap(s), 201, 202, 203, 209, 215

oxidation, 9, 12, 36, 38, 39, 40, 41, 47, 48, 49, 53, 55, 60, 61, 62, 63, 65, 66, 178, 195, 196, 197, 198, 199, 202, 207, 210, 211

oxygen, 9, 12, 41, 176, 196, 197, 198, 202, 203, 204, 207, 209, 210, 211, 212, 214, 215, 219, 220

oxygen transmission rate, 215, 219

P

package, 17, 106, 129, 198, 200, 203, 205, 206, 207, 208, 209, 210, 211, 212, 213, 219, 221, 254

packaging, viii, xi, xiii, 2, 7, 13, 14, 15, 33, 41, 193, 194, 195, 196, 197, 198, 199, 200, 201, 202, 203, 204, 205, 206, 207, 208, 209, 210, 211, 212, 213, 214, 215, 216, 219, 220, 221, 222, 223, 253, 254

packaging design, 200

packing, 2, 7, 170, 193, 194, 199, 200, 202, 203, 205, 208, 209, 212, 213, 253

packs, 196, 200, 201, 204, 206, 210, 211, 215, 216

palatability test, 187

Panicum, 153, 156, 165, 167, 168, 170, 171, 172

pantanal, 229, 235, 238

pasture, viii, 127, 128, 129, 130, 131, 138, 139, 140, 143, 145, 151, 161, 165, 166, 167, 173

pasture management, 152, 153, 156, 165, 170

pasture-fed, 130, 162

Index

permeability, 202, 211, 213, 215
pH of the meat, 226, 231
pigment, 196, 197, 198, 199, 204, 207
plastic, 2, 43, 44, 46, 201, 206, 208, 212, 213, 215, 223, 254
plastic bags, 201, 216
poly vinyl chloride (PVC), 201, 202, 209, 213, 215, 218, 219
polyamide, 214, 222
polyethylene, 15, 201, 202, 214, 215, 216, 253
polymers, 213, 214, 219, 223
polyunsaturated fatty acids (PUFAs), 36, 38, 55, 58, 60, 61, 65, 124, 125, 129, 130, 132, 133, 135, 136, 140, 141, 143, 150, 161, 185
polyvinylidene dichloride (PVDC), 214, 215, 216, 217, 218, 219
postmortem evaluation, 228
pouches, 201, 212, 213, 218
preference of consumers, 17
prenatal development, 176
presence of horns, 229
preservation, 12, 83, 153, 160, 163, 194, 210, 219, 220, 222, 223, 254
pre-slaughter handling, 228, 229, 232, 233, 237, 238, 241
product's shelf life, 40
production, viii, xii, xiii, xv, 2, 4, 6, 7, 9, 11, 14, 23, 24, 32, 36, 38, 65, 69, 70, 71, 72, 73, 74, 75, 76, 78, 79, 80, 81, 82, 83, 84, 86, 87, 89, 96, 119, 124, 125, 126, 133, 144, 147, 148, 149, 152, 153, 154, 155, 156, 157, 159, 160, 161, 163, 164, 165, 166, 167, 168, 169, 170, 171, 175, 176, 177, 178, 181, 183, 184, 188, 189, 190, 199, 222, 226, 227, 230, 232, 235, 236, 237, 238, 239, 241, 243, 244, 245, 252, 254
production chain, xii, 7, 11, 24, 69, 70, 71, 72, 73, 74, 75, 76, 78, 80, 82, 83, 87, 188, 226, 227, 237

productivity, 23, 152, 156, 157, 158, 167, 193, 200, 206, 244
protection, 6, 13, 14, 36, 39, 40, 48, 53, 60, 120, 194, 201, 208, 226, 238
protein, 2, 10, 11, 15, 16, 20, 21, 22, 24, 25, 30, 40, 44, 49, 50, 54, 61, 142, 147, 150, 152, 153, 155, 158, 163, 183, 190, 191, 197, 210, 211, 228, 232
public concern, 144
public health, 1, 2, 5, 9, 14, 22, 72
PUFA/SFA ratio, 55, 58, 136
purge, 54, 195, 201, 202, 205, 210

Q

quality items, 94
quantitative research, 100, 107, 108

R

race, 229
ranch, 74, 230, 234
rancidity, 63, 196, 210
reactive, 41, 45, 225, 230
reactivity, 229
registration, 4, 73, 74, 79
regulatory, 2, 23, 27, 63, 70, 81, 82, 181, 186, 200
retort, 201, 212, 218
river, 234, 235, 236, 238
river transportation, 234, 235
road transportation, 234, 235, 236
roasted, 16, 18
rustic animals, 229

S

salt intake, 94, 95, 102, 104, 105, 110, 112, 113, 114, 116, 117
salt reduction, xiii, 94, 103, 117, 119, 121

Index

sample, 2, 17, 41, 44, 45, 47, 52, 58, 59, 60, 70, 76, 77, 129, 234

sanitary control, 24, 77, 82

sanitary inspection, 5, 7, 23, 25, 30, 33

sanitary safety, 79

saturated fat, 20, 36, 37, 38, 55, 125, 162, 185

saturated fatty acids (PUFAs), 36, 37, 38, 55, 58, 60, 124, 125, 129, 130, 132, 133, 135, 136, 140, 141, 143, 162, 185

sausage characteristics, 106

scavengers, 211, 220

seasoning, 12, 107, 108, 109, 110

sensory attributes, 58, 61, 96, 99, 107, 231

sensory characteristics, 36, 38, 39, 40, 49, 98, 99, 147

sensory descriptors, 94

sensory evaluation, 16, 17, 46, 58, 60, 66, 251

sesame, 12, 16

sesame oil, 12

sexual condition, 229, 230

shelf life, 36, 40, 49, 50, 58, 193, 194, 195, 197, 198, 199, 200, 201, 202, 203, 204, 207, 208, 219, 220, 222, 232, 254

shelf-stable, 212

shrink, 202, 203, 204, 205, 208, 214, 216, 217

SISBOV, 7, 27, 70, 71, 72, 73, 74, 75, 76, 77, 78, 79, 80, 81, 82, 83, 84, 85, 86

skin, 194, 204, 206, 207, 216

slaughter, ix, xv, 6, 74, 84, 127, 136, 147, 149, 153, 159, 178, 225, 227, 228, 229, 231, 232, 233, 234, 235, 236, 237, 238, 239, 240, 241, 242, 244, 245

slaughterhouses, 2, 23, 70, 75, 76, 77, 80, 81, 83, 87, 88, 153, 159, 185, 201, 230, 235, 240, 241

sodium, viii, 2, 10, 12, 20, 21, 22, 24, 36, 37, 40, 42, 43, 49, 55, 93, 94, 95, 97, 98, 99, 101, 102, 105, 108, 109, 111, 112,

113, 114, 115, 116, 117, 118, 119, 120, 253

sodium chloride, 40, 43, 93, 95, 96, 99, 102, 111, 112, 113, 114, 115, 118

sodium reduction, 94, 95, 97, 98, 114, 118, 120

specific cholesterol, 144

spoilage, 196, 202, 203, 211

steel cans, 212

STEPS, 226, 236, 238

storage life, 198

stress, 62, 226, 228, 232, 233, 234, 236, 237, 239, 240, 242, 245, 252

stress agent, 228

stress factors, 232

stressful handling, 227

stretch film, 201, 215

substrate, 126, 141, 186

sunk costs, 80

supply, vii, viii, xiii, 1, 2, 5, 7, 8, 22, 23, 24, 26, 27, 28, 29, 30, 32, 36, 61, 69, 70, 73, 78, 81, 82, 84, 85, 86, 87, 89, 152, 155, 163, 193, 199, 226, 238, 253

survey, 70, 76, 114, 120, 187, 245

sustainability, xii, 23, 153, 164, 220

sustainable production, 227

synthesis, xv, 126, 130, 132, 136, 137, 163, 180, 186

T

taste, 2, 17, 18, 95, 118, 231, 232

taurus breed, 230

technology, xi, xii, xiii, 2, 23, 31, 32, 35, 61, 62, 63, 64, 65, 66, 72, 79, 87, 89, 93, 115, 116, 120, 151, 170, 188, 202, 205, 220, 221, 222, 223, 225, 241, 245, 247, 249, 250, 251, 252, 253, 254

temperature, 12, 13, 16, 18, 43, 44, 45, 49, 100, 128, 185, 195, 196, 199, 200, 203, 207, 211, 213, 214, 221, 233

Index

tenderness, 94, 107, 108, 109, 110, 115, 176, 183, 184, 185, 187, 188, 190, 191, 194, 195, 199, 202, 204, 210, 231, 241
territorial extension, 235
texture/softness, 18
texturized soy, 2, 11, 15, 16, 25, 30
thrombogenic, 124, 129, 137
tolerable upper intake level (UL), 21
total cholesterol, 129, 144
traceability, viii, xiii, 7, 33, 69, 70, 72, 73, 74, 75, 79, 82, 83, 85, 86, 87, 89, 90, 220
traceability system, 70, 72, 75, 87, 90
trade balance, 82
trading bloc, 73, 75
trained, 83, 226, 237, 238, 250
trans-11, 126, 162, 163
transformation, 232
transparency, 15, 80, 83, 86, 214
transportation, 14, 89, 196, 198, 203, 210, 211, 229, 230, 232, 233, 234, 235, 236, 239, 241, 243
transportation conditions, 234, 235
transportation distances, 230, 236
trauma, 229, 237
tray, 201, 202, 206, 207, 209, 213, 218
truck, 233, 236

U

unloading stages, 229

utero-placental blood flow, 176

V

vacuum pack, 194, 196, 197, 198, 201, 204, 205, 206, 208, 209, 210, 215
vacuum packaging, 197, 198, 204, 208, 209, 210
vacuum skin pack, 194, 207
value chain, 200, 220
variability, 124, 129, 145, 146, 206
veterinarians, 77
veterinary medicine, 6

W

weak flesh, 75, 81
welfare, ix, xiii, 225, 226, 227, 228, 229, 235, 237, 238, 239, 240, 241, 242, 243, 244, 245, 254
wholesale, 200, 201, 204, 210
wrap, 198, 203

Z

zebu animals, 187
zebu breeds, 229, 241
zebu genotype, 187